Mechanical evaluation strategies for plastics

Mechanical evaluation strategies for plastics

D R Moore and S Turner

CRC Press
Boca Raton Boston New York Washington, DC

WOODHEAD PUBLISHING LIMITED

Cambridge England

Published by Woodhead Publishing Limited, Abington Hall, Abington
Cambridge CB1 6AH, England
www.woodhead-publishing.com

Published in North and South America by CRC Press LLC,
2000 Corporate Blvd, NW
Boca Raton FL 33431, USA

First published 2001, Woodhead Publishing Ltd and CRC Press LLC
© 2001, Woodhead Publishing Ltd
The authors have asserted their moral rights.

British Library Cataloguing in Publication Data
A catalogue record for this book is available from the British Library.

Library of Congress Cataloging in Publication Data
A catalog record for this book is available from the Library of Congress.

Woodhead Publishing ISBN 1 85573 379 X
CRC Press ISBN 0-8493-0842-9
CRC Press order number: WP0842

Cover design by The ColourStudio
Typeset by Best-set Typesetter Ltd., Hong Kong
Printed by TJ International Ltd, Cornwall, England

Contents

Preface

This book examines strategies for experimental approaches to stiffness, strength and toughness testing. There are numerous options in these activities and one of the aims of the book is to critically consider beneficial approaches in the context of multiple objectives. How test results are interpreted may be more important than the physical validity of the test procedure in some instances. A good test strategy can compensate for imperfections or insufficiencies in a particular procedure and integrate the results from individual test components into an entity that is more informative than the simple sum of the individual parts. Thus, whilst various classes of test have to be segregated for the convenience of a reader, they cannot be so segregated in testing practice because one class of test may support another via correlations that might surprise or offend the purist, but that are invaluable at the sharp end of commercial competitiveness.

The book divides mechanical evaluation into two areas: modulus measurement and strength/ductility measurement. Each area is then considered in terms of the type of input function, namely constant deformation rate, sinusoidal excitation, step-function tests and impulse excitation. Consequently, all of the usual procedures in mechanical property evaluations are accommodated; universal test machine tests, creep modulus, dynamical mechanical modulus, stiffness anisotropy, general fracture toughness, long-term strength, strength in aggressive environments, fatigue and impact toughness.

The presentational style of the book is novel. The 11 chapters dealing with modulus and strength/ductility are quite short. They provide a condensed opinion of what is beneficial and what is mythical, but largely without formal review since much of that is covered in other texts. Consequently, there is no attempt at citing detailed references nor is there a formal account of the many test methods. However, there is a copious section on other recommended reading.

Each chapter contains small but numbered sections. Throughout these sections there are opportunities to provide illustration of a point. These illustrations appear as referenced supplements at the end of each chapter, e.g. Supplement 1.4 discusses samples, specimens and tests. By placing them as appendices, each chapter can be read without deviation in order to obtain an overall view on what is a vast topic. On the other hand, the supplements (which exceed 50 in number) can be a source of specific study in using the book, each one standing essentially self-contained.

The book is aimed at a broad audience of material scientists and engineers. For those in industry, whether involved as material suppliers, processors or end-users, there is an account of the approaches that they can use in characterizing mechanical properties as well as utilizing mechanical properties in end-product application or in contemplating the influence of processing on properties. Students in engineering, materials science or polymer science departments will benefit from the short, yet broad-ranging, chapter content, but also be guided by the many illustrations of application of a principle or an approach.

Both authors have worked in the Mechanical Property Group in ICI's Research Department. They overlapped for 10 years but covered a total span of 50 years from which much of their knowledge and experience has developed. During this long period they had the benefit of working with a large number of colleagues. This was a rich experience, and both authors acknowledge the pleasure, benefits and learning that emerged from the constructive interactions.

Abbreviations

ABS	acrylonitrile butadiene styrene
APC	aromatic polymer composite
ASTM	American Society for Testing and Materials
BPDA	3, 4, 3′, 4′ biphenyl tetracarboxylic dianhydride
BSI	British Standards Institute
BTL	Bell Telephone Laboratory
CLS	cracked lap shear
COD	crack opening displacement
CT	compact tension
c.v.	coefficient of variation
DCB	double cantilever beam
DDE	diamino diphenyl ether
DIN	Deutsches Institut für Normung
DMA	dynamical mechanical analyser
ENF	end-notched flexure
ESC	environmental stress cracking
ESIS	European Structural Integrity Society
HD	high density as in high density polyethylene
HDT	heat distortion temperature
HDPE	high density polyethylene
ISO	International Organization for Standardization
LEFM	linear elastic fracture mechanics
MEK	methyl ethyl ketone
MFI	melt flow index
NDT	non-destructive testing
NPL	National Physical Laboratory
PEEK	poly(ether ether ketone)
PES	poly(ethersulphone)
PMDA	pyromellitic dianhydride
PMMA	poly(methyl methacrylate)
PP	polypropylene

PTFE   poly(tetrafluoroethylene)
PTMT   poly(tetramethylene terephthalate)
PVC   poly(vinyl chloride)
R-curve plot of fracture toughness vs. crack length
SEN   single edge notched
tan δ tangent of the loss angle δ
TC4   Technical Committee 4
3D   three dimensional
w/w   weight/weight concentration

Symbols

a	crack length
a	half-length of intrinsic defect
a	radius of a support ring (re impact equipment)
a	lateral dimension
a_1, a_2	side dimension of a rectangular plate
$a_{0...n}$	coefficients
a_f	effective crack length
A	shape factor
A	cross-sectional area
A	constant
A	coefficient
A_i	a series of coefficients $A_1, A_2, \ldots$
α	highest temperature transition (with respect to loss processes)
b	specimen width
$b_{0...m}$	coefficients
b_{min}	minimum value of specimen width for plane strain conditions
B	coefficient
$B(\sigma, t)$	bulk compliance which is stress and time dependent
β	next to highest temperature transition (re loss processes)
$\beta(b, h)$	shape factor for torsion of a beam
c	$\cos\theta$
c^*	complex compliance for a Voigt element
c', c''	real and imaginary components of compliance associated with a Voigt element
C	stiffness $[E/(1-v^2)]$
$\overline{C}$	transform compliance
$\overline{C}_c(p)$	Laplace creep function
$C^*(\omega)$	complex compliance (dependent on angular frequency)

$C(t)$	time dependent compliance also creep compliance function
$C_c(t)$	creep function
C^*	complex compliance
C', C''	real and imaginary components of complex compliance
$C(\sigma, t)$	uniaxial compliance dependent on stress and time
C_{ijkl}	stiffness coefficients where i, j, k, l can have values of 1, 2 or 3
Cij	stiffness coefficients for an applied stress with the direction of the normal to the surface being i and a strain axis j (where i, j can be 1, 2 or 3)
cv	coefficient of variation
d	tank wall thickness
d	diameter of a cylinder
d	displacement
da/dn	crack growth per cycle
D	flexural rigidity (EI)
D	ductility factor
δ	deflection of a beam
δ	phase angle associated with complex modulus
δ	crack opening displacement
δ_l	increase in length
$\delta(t)$	time dependent deflection
δ_b	deflection due to bending
δs	deflection due to shear
δcrit	critical value of crack opening displacement
Δ	crack length correction
ΔV	change in volume
ΔH	activation energy
ΔT	temperature increment
ΔP	force difference (re fatigue testing)
ΔK	difference in stress field intensity factor (re fatigue testing)
$\Delta \sigma_R$	change in remote stress (re fatigue testing)
e	exponential function
E	axial modulus also known as Young's modulus and as modulus of elasticity
(EI)	flexural rigidity
$E(t)$	time dependent modulus
$E_c(t)$	tensile creep modulus
E_t	modulus at time t
E_c	composite modulus
E_f	fibre modulus
E_m	matrix modulus

E_{ij}	engineering axial modulus where i, j can be 1, 2 or 3
E_r	relaxation modulus in tension
E_θ	in plane modulus where typically $\theta = 0°, 45°, 90°$
$E(100)$	100 sec axial modulus
$E*$	complex modulus
ε_{ij}	strain in the plane i, j where i, j relate to the rectangular co-ordinates x, y, z
ε	strain (engineering)
ε_H	Henky or logarithmic strain
ε_{equiv}	equivalent strain
$\varepsilon_{l,w,h}$	axial, width and thickness strains
ε_{kl}	strain tensor where k, l can be 1, 2 or 3
ε_{ij}	strain tensor where i, j can be 1, 2 or 3
f	frequency
$f(\theta)$	angular distribution of stress
$f(h)$	function of specimen thickness
$f(v)$	function of Poisson's ratio
$f(a)$	function of crack length
$f(\theta)$	function dependent on the angle θ
f_0	constant
f_1, f_2	specific frequencies
$f(\sigma)$	function of stress
$F(\sigma, t)$	function of stress and time
FR	fractional recovery
F	large displacement correction
ϕ	fibre volume fracture
ϕ	geometry function
$\phi(\varepsilon)$	function of strain
ϕ	phase angle associated with complex modulus but used in fixed vibration systems ($\phi \neq \delta$)
g	acceleration due to gravity
$g(t)$	function of time
G	shear modulus
G	strain energy release rate
G	surface energy
$G*$	complex shear modulus
$G(t)$	time-dependent shear modulus
G', G''	real and imaginary components of a complex shear modulus
G_{ij}	engineering shear modulus where i, j can be 1, 2 or 3
$G(100)$	100s shear modulus
G_c	critical value for strain energy release rate but more usually known as fracture toughness
G_{Ic}	opening mode fracture toughness

γ	fracture surface energy
γ	third from highest temperature transition (re loss processes)
γ_{ij}	shear strain where i, j can be 1, 2 or 3

H	height of liquid in a cylindrical tank
h	specimen depth, thickness
h	specimen diameter
$h_{1,2}$	outer and inner specimen diameters, respectively, of a tube
η	viscosity associated with a dashpot
η_α	inefficiency factor for imperfectly aligned continuous fibres
η_c	inefficiency factor for imperfectly discontinuous fibres
η_d	inefficiency factor for debonded fibres

I	area moment of inertia

J_{2D}	second invariant of the deviatoric part of the stress tensor
J_1	first invariant of the stress tensor
J_c	crack resistance

k	stress concentration
k	spring constant
$k*$	complex stiffness
k', k''	real and imaginary components of stiffness for a Voigt element
K	bulk modulus
K_a	apparent bulk modulus
$K*$	complex bulk modulus
K', K''	real and imaginary components of a complex bulk modulus
K^l	Larsen–Miller creep rupture parameter for equating time and temperature
$K_{\text{I,II,III}}$	opening mode, shear mode and antishear mode (tearing mode) for stress field intensity factor, respectively
K_c	critical value of stress field intensity factor but better known as fracture toughness
K_{Ic}	opening mode fracture toughness
K_{c1}	plane strain fracture toughness
K_{c2}	plane stress fracture toughness

l	gauge length
l	length of a cylinder
l	decay length (in St Venat's principle)
l	specimen height
l_o	original gauge length

L	effective beam length or span
L	length of a diagonal
λ	a specific decay length (in St Venat's principle)
λ	extension ratio
m	Paris law integer
m	mass of impactor
m	an integer in the range $0 \ldots m$
$\overline{M}_r(p)$	Laplace creep modulus function
$M_r(t)$	relaxation creep modulus function
M^*	complex modulus
M', M''	real and imaginary components of complex modulus
M_o	instantaneous modulus
M_t	applied torque
μ	coefficient of friction
n	an integer in the range $0 \ldots n$
n	$\sin \theta$
n	power law coefficient
N or n	number of fatigue cycles
N	loading block correction
ν	lateral contraction ratio also known as Poisson's ratio
$\nu(t)$	time-dependent lateral contraction ratio
ν', ν''	real and imaginary components of a complex lateral contraction ratio
ν^*	complex lateral contraction ratio
$\nu(\sigma, t)$	stress and time-dependent lateral contraction ratio
ν_{ij}	engineering lateral contraction ratio where i, j can be 1, 2 or 3
p	a variable
P	applied force
P	peak force
P_{max}	maximum force
P_{min}	minimum force
Q	transverse shear stiffness
θ	an angle
θ	angle measured anticlockwise from the line of a crack
r	a radius
r	an integer from 0 to r

r	distance from crack tip
r_P	plastics zone radius also length of the line zone in a Dugdale model
r_o'	effective area of contact between impactor nose and specimen
r_0	actual area of contact between impactor nose and specimen
R	gas constant (8.314 J/mol K)
R	ratio of minimum to maximum force in fatigue
ρ	density
ρ	notch-tip radius
s	$\sin\theta$
$S.$	span
S	stress
S_{ijkl}	are compliance coefficients where i, j, k, l can be 1, 2 or 3
S_{ij}	are compliance coefficients for an applied strain with the direction of the normal to the surface being i and the line of action of the stress being j (where i, j can be 1, 2 or 3)
σ	stress (including true stress)
σ_o	a step function of stress
$\sigma(t)$	time-dependent stress
σ_a	applied stress
σ_{ij}	stress tensor where i, j can be 1, 2 or 3 (also for σ_{kl}) or x, y or z
$\sigma_{11}, \sigma_{22}, \sigma_{33}$	normal stresses
$\sigma_{12}, \sigma_{23}, \sigma_{31}$	shear stresses
$\sigma_L(r, \theta)$	local stress at the tip of a crack
σ_R	applied stress at a point remote from the crack (notch)
σ_y	yield stress
σi	principal stress where i can be 1, 2 or 3
σ_n or σ_{net}	net stress
σ_{max}	maximum stress
σ_f	stress at fracture
σ_m	stress in the tensile skin layer at mid-span (re bending of a beam)
σ_t, σ_c	tensile and compressive stress, respectively
σ_D	design stress
t	time
t_R	reduced time (re creep recovery)
t_{min}	minimum creep rupture lifetime
t_{max}	maximum creep rupture lifetime
t_{ON}	time on load

T	temperature
T_b	embrittlement temperature
T_1, T_2	specific temperatures
T_g	glass–rubber transition temperature
τ	a relaxation time; for a spring-dashpot array, it is the ratio of viscosity to spring constant i.e η/κ
τ_{ij}	shear stress where i, j can be 1, 2 or 3 or alternatively x, y or z
τ_{max}	maximum shear stress
u	a variable
U	stored energy
Ut	energy stored by a specimen at any time t
v	velocity
v_c	velocity of crack propagation in glass
v_t	velocity of crack propagation in an infinite medium
v_t	velocity of an impactor (re instrumented impact equipment)
vo	initial velocity of an impactor
V	volume
$\Delta V/V$	volume strain
w	width
w_P	energy per unit volume
W	specimen width
ω	angular frequency
x	a variable
x_t	displacement of specimen at any time t
Y	geometry factor in fracture mechanics
z	distance from the neutral axis (for a beam specimen)

1.1

Plastics are polymeric materials and, to varying degrees, are non-linear viscoelastic. Consequently, their mechanical properties are functions of the ambient temperature, testing rate, magnitude of the excitation and other variables. This has strongly influenced the form and nature of the mechanical evaluation strategies and test procedures that have been adopted. Satisfactory performance of plastics in service frequently depends on their behaviour when subjected to external forces and this is reflected in a multiplicity of test methods for the measurement of properties or attributes related to modulus, strength and ductility. Operations such as the development of new materials, quality assurance and the design of end-products, rely at least partly on these so-called primary mechanical properties which therefore have to be mapped in considerable detail.

Other mechanical properties can be classified as secondary in that they impinge on the performance of only some products in some applications and consequently they have not been studied to the same degree as the primary properties have or at a comparable level of generality. Hardness, frictional characteristics and wear resistance are particular examples of secondary mechanical properties, see Supplement S1.1.

1.2

In general, the established test methods are now mainly satisfactory, even though the rapid growth of the plastics industry in the 1950s and 1960s, coupled with an emerging recognition at that time that the mechanical characteristics of plastics were more complex than had been originally thought, engendered the adoption of many classes of test, for example, stress relaxation, creep, creep rupture, with others. The current satisfactory test methods and procedures evolved largely by processes of trial and error but are nevertheless based on the well-tried principles of applied mechanics and a rationale has emerged to provide a framework against which plastics can be evaluated and understood. That rationale uses a mathematical formalism through which any particular specimen geometry in its test configuration can be seen as one of a linked set, each member of which responds in some specific way to the excitation imposed on it. That nomenclature (excitation and response) usefully incorporates time into the stress–strain relationships thereby embracing the viscoelastic character of the materials.

When the response of a test specimen is linear with respect to the excitation, or otherwise regular, a property can be defined and evaluated in terms of that excitation and response. When it is not linear, the definition may still be valid provided the 'property' is acknowledged as being sensitive to the magnitude of the excitation but a property value derived thereby will be only one of many possible values. In either case, the excitation–response relationship is a function of several variables, the principal ones for a plastics material being the temperature, the testing rate and the magnitude of the excitation.

1.3

The properties of plastics that are related to strength and toughness, e.g. creep rupture lifetimes, fatigue resistance, impact resistance, are less regular than those related to modulus. That is largely because modulus is a relatively simple concept well supported by viscoelasticity theory that provides the requisite extensions to classical elasticity theory, whereas strength and toughness are not simple concepts, do not have the same degree of support by theory, and are prone to the consequences of local fluctuations. Plasticity theory is helpful in some strength and toughness issues and fracture mechanics is helpful in the case of brittle fracture but most service failures involve a combination of pseudo-plasticity and crack growth and are neither simple to record nor tractable analytically. Furthermore, data from destructive tests tend to be more variable than data from modulus tests and exhibit a level of unpredictability that engenders a wider range of tests.

In many cases the test configurations for measurements of strength can be the same as those for modulus measurements and sometimes the specimen geometries can also be the same but the test pieces often incorporate an additional feature to create a stress concentration or simulate a defect. That additional feature may be grossly obvious, for example, a notch, a groove, a crack, a rough surface finish; it may be textural, for example, a knit-line, a phase boundary; it may only be obvious at a molecular level, for example unfavourable orientation or some variation in the degree of cross-link density. The literature abounds with reports of functional relationships and correlations between processing variables, structural features and toughness, but no clear picture has emerged and the subject remains essentially empirical. Even so, there seems to be no supporting case for radical changes to the test methods themselves.

1.4

Despite the supporting theories being simplifications and idealizations of the actual behaviour of plastics, the various mechanical test procedures for such materials have already evolved into what is now a comprehensive and interconnected system of tests, see Fig. 1.4/1. The format of the evaluation strategy that is described in the chapters that follow has been dictated largely by the viscoelasticity and, though many of the test methods and procedures evolved into their current forms through considerations of experimental convenience, their consolidation nevertheless rests on a pragmatic

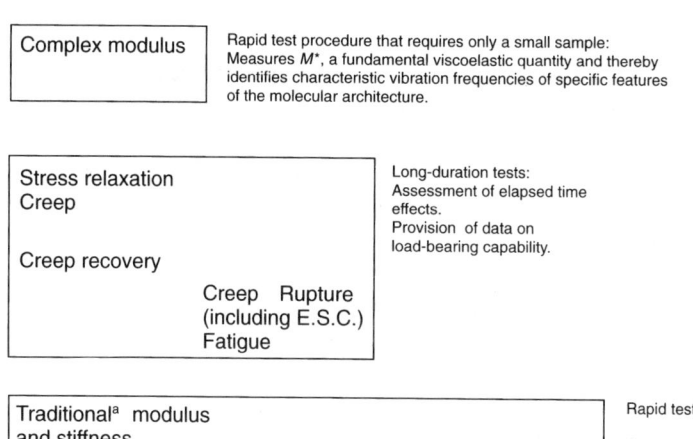

a'Traditional' refers to those tests usually carried out on the commonly misnamed 'tensile test machines' at constant rates of deformation, i.e. ramp-excitation tests (see later).

bThese quantities may be measured during a 'traditional' test or during creep rupture or fatigue experiments depending on circumstances and intended utilization.

1.4/1. System of mechanical tests for the evaluation of plastics materials and the prediction of service performance.

interpretation of viscoelasticity theory. Thus, in what follows, properties are considered mainly in the context of the responses of plastics to specific and essentially simple excitation functions, namely step, ramp, sinusoidal and impulse excitations, with a few others derived from linear superposition of those basic excitations.

Even so, practical complications arise in evaluation programmes because mechanical properties are very sensitive to the nature of the plastic, in particular, whether it is a thermoplastic or a thermosetting material. The sensitivity of the properties relate to preparation and processing characteristics of these different types of plastics. Preparation factors might include network characteristics for thermosetting materials such as phenolic, epoxy and some acrylics polymers, where the cross-linking chemistry which is activated during the preparation of a sample will strongly influence important structural aspects of the material. For thermoplastic materials, preparation factors will include, for example, cooling rate from the melt and its subsequent influence on the level of crystallinity (in materials such as polyethylene, polypropylene, poly-ether-ether-ketone, polyamides, etc.) or molecular packing efficiency (in amorphous glassy polymers such as poly-methymethacrylate, polyethersulfone, polystyrene, polycarbonate, etc.). Similarly, processing factors include flow path length by which the molten polymer is converted into a solid shape. In turn, the flow of a cooling mass of molten polymer into mould cavities or through extrusion dies results in complicated patterns of melt transport and of consequential molecular order in the end-product that, in turn, affect its mechanical properties. The principal consequences are that the various phenomena associated with modulus, strength and ductility vary from point to point in a moulding, vary with test direction at any particular point and vary from batch to batch as the processing conditions are changed. Those features raise fundamental questions about the nature of, and the relationships between, samples of a material, intermediate processed forms of that material, end-products made from that material, test specimens, tests and test data, see Supplement S1.4.

These preparation and processing factors are not commonly or universally relevant for themosetting or thermoplastic materials. However, since this book will relate to mechanical property strategies for both classes of plastics, the detailed aspects and consequences of preparation and processing will be used as and when necessary without qualification.

1.5

The plastics producers, who were the main data generators until recently, originally used either test specimens cut from isotropic sheets (for example, compression moulded sheets for thermoplastics or cured or cast plaques for thermosetting materials) or anisotropic and/or heterogeneous sheets (for example, intact injection-moulded test pieces for thermoplastics or consolidated preimpregated continuous fibre reinforced tapes for thermoset composites). It is now believed that, in general, the so-called 'properties of materials' have little intrinsic value – materials have to be processed into usable forms and it is the properties of those usable forms that are important commercially. However, by the same argument and in the light of Supplement S1.4 the evaluation programmes that apparently served their earlier purposes satisfactorily, judging from the outstanding growth of the plastics industry during the period 1950–1970, are insufficiently comprehensive in relation to modern applications.

1.6

In principle, the variation in property values due to the processing variables and related factors can be accommodated by additional testing; the theory of the mechanics of anisotropic systems stipulates the form of the additional tests, the processing conditions can be varied systematically and correlations can be established between the properties of samples, specimens and end-products. In practice, however, the anisotropy in test specimens and commercial mouldings often turns out to be far more complex than the systems embraced by theory, even for the relatively simple case of modulus, and the critical processing parameters cannot always be varied independently of one another. Thus, a modified evaluation programme that merely expands earlier ones may not necessarily be additionally and usefully informative in proportion to the increased cost and effort.

1.7

A more fundamental issue is that in the light of Section 1.5 and Supplement S1.4 the measured quantities cannot be regarded as physical properties in the true sense of the word and therefore they are not definitive. That being so, the standard constraints on how tests should be conducted and samples evaluated may be unnecessarily restrictive and even inappropriate. Arguments in favour of a revised testing strategy can be advanced thus:

1 preparation and processing factors are critical variables and should be treated as such in evaluation campaigns;
2 there is likely to be a population of possible values of a 'property' which are separately appropriate to particular end-products and service situations;
3 the cost implications of items 1 and 2 might be compensated for by modifications to testing programmes and simplification of some test procedures;

4 the modifications might entail the use of unconventional specimens and test configurations.

1.8

The prospective revisions to the standard concept of test specimens and property data are wide-ranging in that they affect both the overall strategy and the tactics of materials testing activities. The former entails a major change of emphasis (see Section 1.9) in reaction to a growing conviction that the classical conventional approach is becoming progressively less rewarding because of the complexity of the influence that preparation and processing conditions have on the properties of end-products. The prospective change in tactics is correspondingly revolutionary; in essence it entails some reduction in the volume of conventional testing as currently standardized, the adoption of sample state as a variable in the design of experiments and the introduction of tests on possibly unconventional specimens chosen as representative of various important classes of preparation and processing.

1.9

Apart from such *ad hoc* testing, it was usual in the past for polymer scientists to work against the background:

processing → structure → properties

and in the belief that detailed knowledge about the structure would give crucial insights into the mechanical behaviour and greater efficiency in development programmes. This generated a complex description of properties and worries about a 'bottomless pit' of study. However, suggestions for limited approaches could be made on the grounds that:

1 the utilization of a material in an application depends on many complex and interacting factors – the influence of enabling science is often marginal because it cannot be expected to quantify the outcome of the numerous and interacting factors that are influential during the processing stage;
2 the historical record shows that developments have usually been technological, with science providing an explanation or a rationalization later;
3 science pursues mainly a process of analysis, exploitation of materials requires a process of synthesis.

Thus, in reaction to the escalating evaluation costs and the elaboration of the rationale, a trend towards the simpler strategy:

processing → properties

emerged, though the relative simplicity embodied in that progression has since been disturbed by the intrusive importance of molecular order in the glassy state as a factor affecting properties. See Supplement S1.9 for additional notes on the changed strategy.

1.10

In view of the large number of influential variables and their complex interactions it seems likely that the most economic course in evaluating plastics is to regard any measured value of a property such as modulus or strength as merely one from a statistical distribution of many possible values that can be manifest in service; the main controlling parameters, which are the preparation and processing factors, should therefore be accorded the status of experimental variables when that is appropriate or feasible and otherwise should be stated as qualifying accompaniments to the data.

1.11

The use of unconventional specimen shapes stemmed from the recognition that the specimens of simple shape that are the most appropriate on the basis of the theory of mechanics seldom incorporate the preparation and particularly the processing factors that commonly arise in service items and that consequently their properties often differ significantly from those manifest in end-products. For example, many moulded end-products contain several specific flow geometry features so that a single specimen cut from a particular region does not necessarily yield a 'representative' property value on which design decisions such as safe cross-sectional areas for load-bearing functions, should be based. However, those specific flow features can be separately simulated by appropriate moulded shapes, which can serve directly as test specimens or as a source of test specimens for the generation of service-pertinent data. Such specimens have been referrred to as 'critical basic shapes', see Supplement S1.11 and also Sections 2.5 and 2.7 in relation to a specific evaluation issue.

The critical basic shapes envisaged above correspond to the 'structural elements' referred to in engineering texts, i.e. basic structural units such as tension bars, torsion members, box beams, etc. An assembly of several structural elements constitutes a structural component or an entire end-product in which each element exhibits unique property values corresponding to those derivable from tests on the critical basic shapes. Therefore, the latter should be superior to conventional test specimens as sources of data for the engineering design of end-products.

1.12

Irrespective of whether the procedures are conventional or unconventional, tests are carried out for various purposes that separately demand different attributes of the test procedures. The results from mechanical tests on materials are used for several radically different purposes, i.e.

- as criteria in quality control and quality assurance activities;
- as a basis for the comparison and selection of materials;
- as data for design calculations;
- as a basis for predictions of service performance;
- as an indicator in materials development programmes;
- as a starting point for the formulation of theories in materials science.

Those various objectives separately demand different attributes of the test methods. Thus, for quality control and quality assurance purposes it may suffice for the tests to yield reproducible data, whereas design data should be precise, accurate and pertinent to the intended service. In turn, design data may be founded on enlightened empiricism with little direct support from theory other than the long-established theory of mechanics whereas data intended to constitute a basis for the prediction of performance under conditions very different from those of the experimentation should have additional validation from a supporting theory. To serve as indicators in materials development programmes, the test data should be especially sensitive to changes in the influential factors under investigation.

An additional complicating factor is that the differences in the testing objectives of the plastics producers and the manufacturers of plastics end-products engender more than a different testing strategy. In general, the plastics producers are the principal data generators whilst the manufacturers of end-products tend to be data utilizers though the latter group inevitably engage in some testing activities. Inescapably, however, phenomena and properties are interpreted in different contexts by the two groups with the result that the interchange of ideas and the flow of information may be impaired, see Supplement S1.12, though there is an increasing tendency for a primary producer (generator) and a utilizer to collaborate in the development of a new end-product and the information pathways between the participating parties are then usually efficient.

1.13

The prediction of likely service performance on the basis of laboratory experiments falls into at least five distinct categories:

- prediction of the future course of a progressive change that has been quantified only in its early stages (the elapsed time effect);

- prediction of the performance under loading conditions very different from those studied in the laboratory (the arbitrary excitation effect);
- prediction of the performance of articles of complex shape from data derived from specimens of simple geometry (the geometry effect and the stress geometry effect);
- predictions of the behaviour of one sample from data derived from another sample produced under different conditions (the processing or preparation effect);
- predictions of the 'quality' of a sample or estimates of its properties from a small set of test data.

Those prediction processes rely variously on standard analysis in the fields of elasticity, plasticity and viscoelasticity and on procedures such as extrapolation, superposition, accelerated testing, statistical analysis and mathematical modelling. It is important that the path from the original data to information as utilized should be fully documented and traceable. For instance, a utilizer should always know whether a datum was measured directly or estimated via one of the procdures listed above. Similarly, he/she should be given information on the reliability of a datum (but see also Section 1.15).

1.14

The mechanical testing of plastics rests on an elaborate infrastructure of knowledge and procedures that are used in varying combinations depending on circumstances but under the general grouping:

- enabling theories;
- practical execution;
- interpretation of results.

The relevant enabling theories are:

- the mechanics of elastic and viscoelastic solids and of viscous liquids;
- aspects of polymer science and physics;
- experimental design and statistical analysis.

The practical execution entails:

- choice of appropriate sample(s) and test specimen(s), with regard to the purpose of the tests;
- preparation of specimens;
- exploratory experimentation;
- design of experiments;
- standardized and/or *ad hoc* testing;
- experimental programmes.

The interpretation of experimental results entails:

- data processing;
- analysis of results and assessment of errors;
- comparisons of new data with archival data;
- data presentation.

That wide range of operations requires the deployment of a correspondingly wide range of skills which is rarely fully covered within the confines of a single academic discipline or the expertise of a single individual.

1.15

The practical details of many of the tests that are used are specified in Company, National or International Standard Methods of Test, see Supplement S1.15. The analysis of results often entails recourse to statistical analysis because the combined effect of numerous influential variables can distort test data but there is often a conflict of interests therein between a desire for economies in testing costs and the requirement for adequate set sizes. For example, five specimens are often stipulated (as a minimum) for replicate tests, even though that provides only poor estimates of the mean value and the standard deviation when the data are scattered. Additionally, if the data are closely similar apart from one outlier there is some uncertainty as to whether that outlier should be disregarded; rejection of a datum solely on the basis of the standard statistical criterion may be unsafe but if there is independent evidence, such as imperfections in the test piece, the outlier datum should be replaced with a value equal to the lowest (or the highest as the case may be) in the remaining set. Guidance on the subject of the reliability of mechanical test data is given in an NPL publication following a collaboration between several European organizations.[1]

1.16

The interpretation of experimental results ranges in scope from the minutae of the precision and accuracy of particular data to general inferences about other samples of the same material. The main sources of unreliability in mechanical test data are:

1   uncoordinated and/or loosely specified samples;
2   atypical or defective specimens taken from the sample;
3   faulty experimental design;
4   test configuration or specimen such that the stress (or strain) field in the gauge section differs from that intended, expected or assumed;
5   inadequate, inappropriate or badly sited sensors;

6   inadequate control of test conditions;

7   faulty or otherwise inadequate data processing.

Nowadays, many of those sources of unreliability can be eliminated, or at least restricted, by adherence to operating rules and constraints, but materials-related sources are less easily controlled because of the pervasive influence of the production path converting material into product (end-product or testpiece), which can restrict the freedom of choice of the experimenter. That can be an important element if the provider of the test sample is not closely attuned to the objectives of the investigator (academic investigators should be particularly wary of gratuitous samples obtained from a possibly disinterested supplier). Under modern guidelines the sample should be described in sufficient detail for its quality and fitness-for-purpose to be gauged.

1.17

The impediments to precision, accuracy and pertinence are greater for polymeric composites because of the heterogeneity and often an associated anisotropy. Composites with a polymeric matrix include the following:

- plastics with an admixture of rubbery particles;
- plastics with an admixture of mineral particles;
- plastics with an admixture of discontinuous fibres;
- plastics with continous fibres;
- plastics with more than one class of included phase;
- laminates with layers of different materials;

and one might even regard a crystalline polymer as a composite because the crystalline phase is embedded in an amorphous matrix. Some laminates, notably packaging materials, are not commonly regarded as composites.

   The overall mechanical behaviour of composites is affected by the properties of the separate phases, by the spatial disposition, aspect ratio and uniformity of dispersion of the second phase and by the character of the interfaces. Such materials being heterogeneous, it is doubtful in some cases whether the laws of continuum mechanics, e.g. the theory of linear viscoelasticity, are valid. Non-spherical inclusions induce anisotropy, particular practical consequences of which are:

1   severe 'end-effects' which extend in the direction of highest stiffness (a function of both the specimen geometry and the anisotropy);

2   unexpected failure in the grips and at other loading points;

3   delamination and other unintended failure modes;

4   property imbalances between, say, a tensile modulus (or strength) and a shear modulus (or strength); such property imbalances induce behav-

iour of end-products and test coupons that is peculiar to anisotropic systems and also entail restrictions on test configurations and specimen geometries.

Erroneous data can arise also from imperfections in fabricated end-products (including test pieces). The possible imperfections in fibre composites include misalignment of the fibres, poor dispersion, poor coupling between the phases, degraded fibres, and included gases; the imperfections in plastics containing particulate fillers include agglomerates and inadequate coupling between the phases (strong coupling is essential for some purposes, weak coupling for others).

1.18

Many of the points mentioned in preceding sections are relevant to testing and experimentation in general but some relate uniquely to plastics, which are a subclass of viscoelastic materials. The form of the evaluation strategy that is described in the chapters that follow has been dictated almost entirely by that viscoelasticity. Thus, in what follows, the mechanical properties of plastics are considered mainly in the context of the responses to specific and essentially simple excitations, namely steps, ramps, sinusoids and impulses. There was, however, a final twist in the evolutionary development of the current strategy. Most of the critical facts relating to the mechanical behaviour of plastics had been established by the mid-1980s, and genuinely exploratory experimentation was no longer strictly necessary. The time had come for the strategy and the test procedures to be consolidated and, because of increasing commercial pressures, to be made cost effective and pertinent to industry's needs. That process has not yet been completed but this monograph seeks to reflect that change by emphasis on the activity of mechanical testing rather than on the latest and/or elaborate variants of the actual test methods, except where the latter confer strategic benefits that would otherwise be unattainable. Thus, for example, Chapter 2 sets out how relatively crude measurements of two properties can provide valuable insights into the nature of a material whereas, in contrast, Chapters 6 and 10, among others, demonstrate that high accuracy and precision are essential for some purposes.

Reference

1 Various. '*Manual of Codes of Practice for the Determination of Uncertainties in Mechanical Tests.*' N.P.L. Publication, 2000.

S1.1   Secondary mechanical properties: hardness, friction characteristics and wear resistance

The hardness, frictional characteristics and wear resistance of a plastics material are sometimes important and even vital for the satisfactory service performance of an end-product but are not as universally important as modulus, strength and ductility, to which they are related. To varying degrees, they can be classed as surface properties whereas modulus, strength and ductility are bulk properties.

S1.1.1 Hardness

Hardness is less of a surface property than the others. It is related to the yield stress and to the viscoelastic characteristics at lower stresses; it is measured in terms of the resistance to penetration by a loaded indentor. Hardness tests are ill-defined in that they involve a combination of elastic and plastic deformation. The development of a plastic zone beneath an indenter is revealed by the formation of a zone of anomalous refractive index and by the shape of the indentation. Cracks develop with increasing load and the pattern of fracture delineates the directions of the principal stresses.

There is no satisfactory natural standard of 'hardness' and there are more than 20 different scales of hardness. The first tests were merely comparison tests but standard specifications were drawn up later. Commercial hardness testers are now checked with test blocks that have been calibrated against a standard machine, which provides an accuracy of about 2% for an industrial tester and 0.5% for a standard machine. Generally the Brinell and the Rockwell tests are used for the more rigid plastics and the Shore test is used for the softer ones. The first two use spherical indentors and the last one uses what is essentially a needle with either a spherical tip or a truncated conical tip. In all three the penetration is time dependent, as would be expected from the nature of the primary properties, and all three are prone to errors due to friction, which may reduce the indentation pressure by up to 30%. With conical indenters, the indentation pressure varies with cone angle, and adhesion of material to the indenter face causes a cap of dead material to form ahead of cones of angle greater than 60°; in deep punching the rigid movement of that cap is accommodated by zones of intense shear and the dead material may detach from the specimen during the unloading.

Hardness tests on plastics have mainly been somewhat arbitrary and resorted to only when the commercial success of an application depends on the hardness. However, there have been fundamental studies that have related hardness to structure. Thus, Bowman and Bevis[1] assessed the Vickers microhardness test in some detail and related their results to structural features in their samples and Kent et al.[2] used ball indentation to study cracking and molecular orientation. They indented plastics such as polystyrene and polymethylmethacrylate with ball-bearings and showed that the associated radial cracking follows the direction typical of the surface orientation in the moulding. Moreover, the load required to produce the first crack indicates the degree of orientation (findings corroborated by optical birefringence studies). The energy theory of brittle fracture shows that the radial cracking that accompanies plastic–elastic indentation only starts when the plastic zone has attained a certain size; thus, for a series of geometrically similar indentations there is a critical radius below which no cracks form.

In other experiments, injection-moulded, double-gated, opposing flow discs of poly-ethersulphone (PES) did not crack radially at an indentation speed of 1 mm min⁻¹ at room temperature. The only cracking observed was with balls of diameter large compared to the disc thickness (e.g. diameter = 17 mm, thickness 3.5 mm), when the penetration was comparable with the thickness of the specimen; cracking occurred as a star-like or flower-like formation between the flat back of the specimen and bottom of the indentation but without distinguishable orientation. A multitude of intersecting slip-lines were also produced around such indentations. At liquid nitrogen temperature (−195 °C) and with the same test configuration, the moulding shattered from the indentation, with radial cracks running out in all directions plus a ring-crack around the site of indentation (e.g. diameter = 17 mm, thickness 6.7 mm, penetration 5.94 mm and load 7200 kgf at shattering). A cone indenter splits a knit-line open at room temperature with the crack lying along the knit-line: a star-like crack is found in the body of the material.

When a moulding was being indented a rhythmical clicking sound was often heard, accompanied by a drop in the load, which subsequently recovered; no radial cracking was found in the specimens. The clicking is attributed to energy being expended in three processes:

1   friction between the surfaces of the ball and specimen or compression table and specimen;
2   production of slip lines within the specimen;
3   expansion and contraction within the test assembly.

Cracking of the moulding was accompanied by a loud bang and a correspondingly sudden drop in the load.

In the experiments, 52 indentation tests produced 18 radially cracked specimens with varying amounts of deformation and stress whitening. Seven of the 18 were cracked in liquid nitrogen. Unreinforced PES specimens only cracked at an indentation speed of 1 mm min⁻¹ or when they were cooled to –195 °C. Glass fibre reinforced PES specimens cracked at room temperature under the larger balls and smaller balls cracked identical regions of the mouldings at liquid nitrogen temperature. Generally, the more highly filled the specimen the easier it was to crack. Similarly, the gate, jet and knit-line regions were progressively more easily cracked. Radial cracks tended to follow visible defects, e.g. knit-line or jetting, and obvious flow axes, see examples in Fig. S1.1/1.

For each indentation, a crater formed in the surface with a piling up of encircling material and, particularly for filled specimens, the region around the crater became more opaque; both the 'piling-up' and opaqueness were anisotropic in their distribution. The quantitative results were confused by the amount of work expended on plastic deformation, particularly where the indentation and ball radius were comparable or greater than the thickness of the moulding. Nevertheless, the indentation hardness at liquid-nitrogen temperature and the load required to produce cracking are useful quantities.

Similarly, hardness tests (Avery Hardness Tester in accordance with Procedure B of ASTM D 785–65) on a coupled grade and an uncoupled grade of polypropylene containing about 0.02 weight fraction of short glass fibres left virtually no residual deformation on the former but considerable stress whitening (mainly phase separation) in the latter. Also, the hardness decreased in the region of the merging-flow knit-line in double-feed weir-gated mouldings, see Fig. S1.1/2, reflecting the pattern of fibre alignment typical of such mouldings.

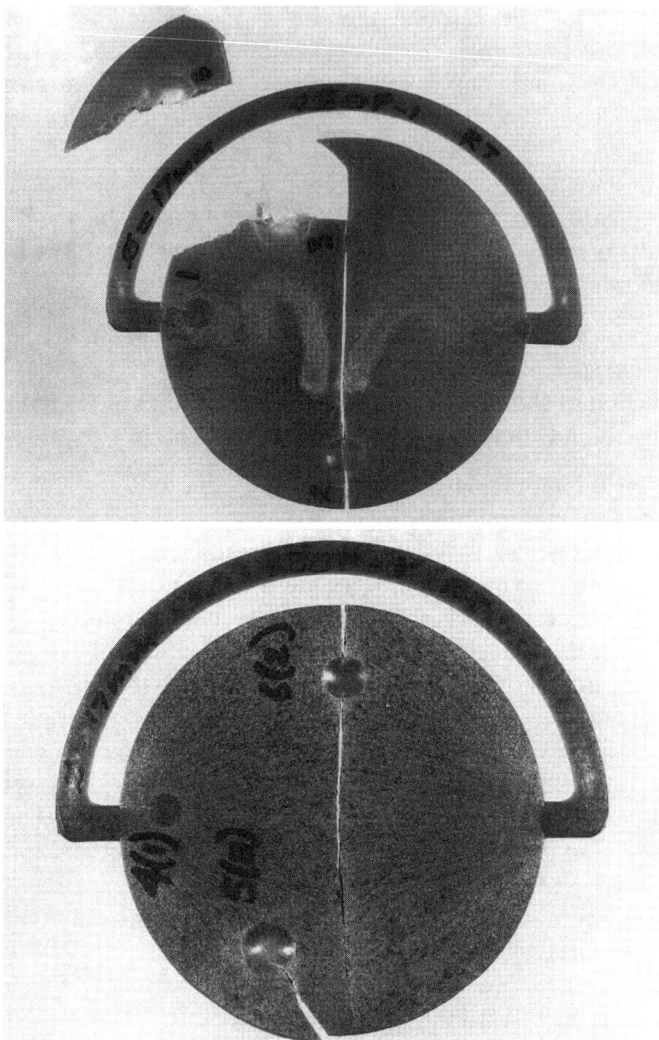

S1.1/1. Indentation and fracture of injection-mouldings of poly(ether sulphone) containing short glass fibres. Fractures along knit-lines and elsewhere. Note 'jetting' flow defect in upper disc.

S1.1.2 Friction

When one surface slides across another, a frictional force arises, partly as a result of adhesion at contacting asperities and subsequent shear rupture of the junctions and partly because the asperities of the harder material plough into the surface of the softer material. The frictional force is sup-

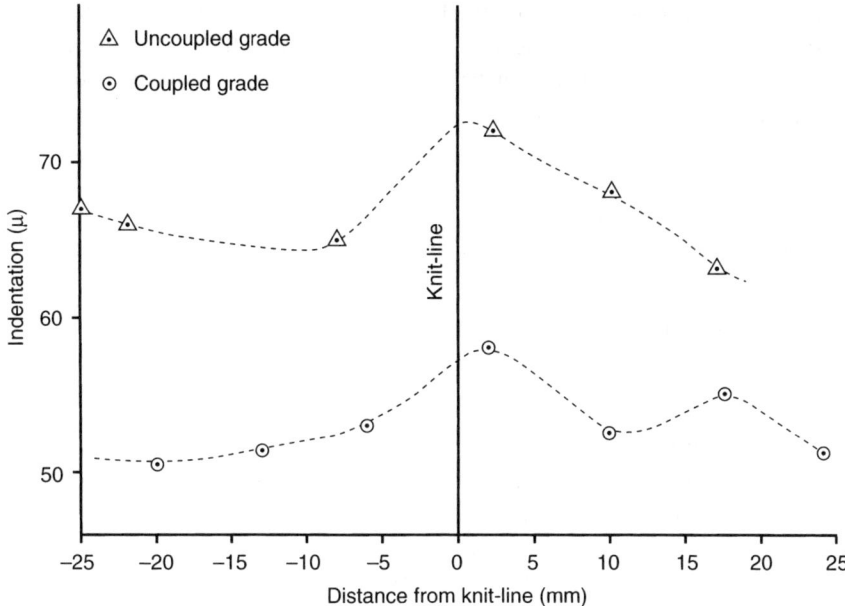

S1.1/2. Hardness of polypropylenes, 0.020 weight fraction of short glass fibres. Hardness decreases in the region of the knit-line and increases with the efficiency of the matrix–fibre coupling.

posedly proportional to the normal force and independent of the apparent area of contact, whence the definition of the coefficient of friction, μ, as:

$$\mu = \frac{\text{Frictional force}}{\text{Normal force}} = \frac{\text{shear strength}}{\text{compressive yield strength}}$$

but here again viscoelasticity, anisotropy and surface texture cause variations in the measured values of μ. It is found that the measured coefficient increases as the normal force decreases because of the increasing preponderance of viscoelastic effects; it increases with increasing temperature for similar reasons but may pass through a maximum in association with a peak in the imaginary part of the complex modulus; it decreases if polymer is transferred, as it often is, to the mating surface during sliding and it increases if hard debris becomes embedded in the polymer surface. It depends also on the smoothness, cleanliness and dryness of the contacting surfaces.

Typical values of the coefficient when one surface is a plastic lie in the range 0.3 to 0.6, though poly tetrafluoroethylene (PTFE) is an exception with values in the range 0.05 to 0.1. However, the effective values in service depend on the presence of lubricants (which in some instances can raise the coefficient), the efficiency with which frictional heat is conducted away, the presence of wear debris at the interface and mechanical details of the item and the service situation. There is a wide range of situations in which the magnitude of the coefficient of friction is important; it is noteworthy that the coefficients of friction of plastics are low enough for several plastics to be suitable for bearings and gear wheels and yet high enough for friction welding to be practicable, and that the friction of an automobile tyre and a road surface can vary from that necessary for efficient traction down to virtually zero, depending on the state of the contacting surfaces.

S1.1.3 Wear resistance

Wear is not directly related to the friction characteristics though the factors that cause an increase in the latter usually cause one in the former. It is normally attributed to four mechanisms: adhesive wear, abrasive wear, corrosive wear and surface fatigue. The first arises from the shearing of the junctions that form between contacting surfaces, the second is caused by entrapped particles, corrosive wear is analogous to environmental stress cracking and surface fatigue is analogous to bulk fatigue. Most of the testing is arbitrary and the results depend on the machine dynamics; for example, in a series of pin and disc wear tests on a 95% alumina sample the wear obtained in a pneumatically hard system was two orders of magnitude greater than that obtained in a pneumatically soft system.[3] Thus, predictions of service performance are best based on protracted tests in devices that approximate to the service situations.

References

1 J. Bowman and M. Bevis. 'The evaluation of the structure and hardness of processed plastics by the Vickers microhardness test.' *Colloid and Polymer Science*, **255**, 954, 1977.
2 R. J. Kent, K. E. Puttick and J. G. Rider. 'Detection of orientation in injection mouldings by indentation fracture – industrial applications'. *Plastics and Rubber: Processing and Applications*, **1**, (2), 111, 1981.
3 M. G. Gee. 'Wear metrology; the art of determining a material's performance.' *Materials World*, **1**, (5), 281, 1993.

S1.4   Samples, specimens and tests

A mechanical test is normally carried out on a specimen taken from a sample of a material, where the sample is either a fabricated item from which the test specimen is machined or a batch of nominally identical, directly moulded specimens from which individuals are selected. The mechanical properties of a thermoplastics sample are sensitive to the processing route and to the subsequent storage conditions. Also, the macroscopic quality of the pieces influences the outcome of a test. Directly moulded test pieces may contain voids and other internal defects, they may subsequently distort, nominally rectangular cross-sections are likely to be trapezoidal, the surfaces may not be flat and they may suffer from a range of surface imperfections. Specimens machined from a sheet or block of material may distort subsequently as the pattern of internal stress changes and they may have surface defects caused by an inappropriate cutting speed or cut direction. Production routes for test specimens each have their particular advantages and the best choice depends on the purpose of the test and the particular circumstances but the additional cost of a machining stage may often outweigh technical considerations. There are analogous considerations in relation to thermosetting resins and composites, with corresponding compromises necessary between sample quality, typicality and cost.

If the tests are intended to provide general characterizing data or data for use in design calculations, it is probably preferable that the molecular state of the sample should be similar to that of typical end-products made from the material. At the very least, this requires that the processing route and the conditions should be similar to those for the production of service items, but this is not necessarily achievable, for example if the production item is very large or very small. If the test programme is intended to establish correlations between processing conditions and properties, the influential factors have to be varied methodically and allowance made for interactions between those factors. Three issues usually have to be resolved, namely:

1   Do the data properly represent the property in view of the particular circumstances of the test (excitation type, test configuration, state of molecular order, etc.)?
2   Do the data demonstrate a unique relationship between the property value(s) and the independent variables?
3   Is there evidence of a genuine cause–effect relationship or is it merely a correlation?

The relationships between processing conditions and properties are such that firm answers can seldom be given to the second and third questions, an additional complicating factor being inadvertent coupling between the control parameters. Partly for that reason, compression mouldings tend to be favoured as sources of test pieces for scientific studies because they can usually be obtained with uniform and reproducible molecular texture and low levels of residual strain which is helpful at the interpretation stage of the work. However, since the cooling rates attainable during compression moulding are generally lower than those attainable during injection moulding and the molecular structures correspondingly denser the mechanical property values derived in such studies tend to differ from those manifest in service. That may be unimportant in some circumstances, but a relationship established for one molecular state may not be similarly valid for a different molecular state; the property values themselves should not be used unconditionally in, say, an engineering design calculation. A concise summary of the range of microstructural features that arise in thermoplastics is given in reference 1.

The molecular state and the mechanical properties of any sample, irrespective of the initial production method, can be modified by an annealing or a conditioning treatment. At one time those procedures were simple (for example the standard annealing treatment for polyethylene was a short period immersed in boiling water) but they progressively became more elaborate as the complexities of the relationships between structure and properties were systematically exposed. Thus, for instance, increased crystallinity generally confers higher modulus, higher yield strength and reduced ductility but the size of the crystals and their agglomeration in spherulites and other entities also affects those properties. The first insight into those complexities was gained through protracted studies of the behaviour of the polyethylenes during the 1950s and 1960s and the subsequent years have seen further refinements of the consensus, with consequential implications for the design of experiments. Similarly, the effects of a high temperature conditioning period on the properties of amorphous plastics, which are attributable to a reduction of the free volume, are much more complex than was originally thought.

Annealing and conditioning procedures put samples and/or specimens into particular states of order which influence any subsequently generated properties data. Therefore, experimental results and quoted property data should not be divorced from the storage history of the test specimens. Major effects on property values, e.g. a twofold change in the modulus of a polyolefin plastic and a four-decade change in the creep rupture lifetime of a PVC, have been identified as attributable to annealing or conditioning treatments and therefore pretreatment, sample, specimen and test procedure should be regarded as a single entity and no property datum in isolation should be regarded as a unique characterizing quantity.

That seemingly unavoidable impasse has raised fundamental questions about the nature of testing, the validity of established test practices, the usefulness of any one set of data, and the shortcomings of current databases and their future scope. It almost seems that tests to compare the mechanical properties of competing materials, to optimize a product or a processing procedure, to generate design data, etc. are inevitably fallible to some degree because of that pervasive influence of molecular state (and other factors) on the properties.

The dilemma may be resolvable by the adoption of a testing strategy differing from the classical one, for instance that outlined in Supplements S1.11 and S1.12 and in other chapters.

Reference

1 D. Hemsley. 'The importance of microstructure to the in-service survival of plastics products.' Guest Editorial, *Plastics, Rubber and Composites: Processing and Applications*, **16**, 2, 77, 1991.

S1.9 Evolving evaluation strategies for thermoplastics

With some simplification, the 1950s may be seen as the period during which the main structure/property interrelationships were identified, the early 1960s were the period during which attempts were made to quantify them, the 1970s were the period during which the recognition of the essentially unbounded nature of that task prompted a cautious return to directed empiricsm and the 1980s were the period during which properties data were increasingly assembled in databases and subjected to numerical manipulation in computer-aided design calculations.

During those periods of development and change, the prevailing evaluation strategies changed correspondingly, taking account of new knowledge, new downstream applications and changing market opportunities. Current strategies are also likely to change in the future. The plural word, 'strategies', is correct because the strategy should depend on the end purpose being served and, at any one time, a primary producer, for instance, is likely to have objectives that differ from those of an academic researcher. On the other hand, it would be advantageous for all concerned if there were to be a single, comprehensive, all-purpose strategy, parts of which could be adopted, and adapted as necessary, to serve the particular purposes of primary producers, designers, academic researchers, end-users, etc.

An empiricist can argue that our present understanding of the relationships between structure and properties is as deep as can ever be useful; the main features of the mechanical behaviour can be interpreted qualitatively in terms of structure and thermodynamic quantities and correlated with the processing conditions. It has been argued, in favour of more research on the structure–property relationships, that a comprehensive knowledge of the structure would lead to the understanding that alone could prevent a proliferation of test methods and data, but that line of reasoning is questionable on two grounds. It assumes, firstly, that the enabling science could be expanded sufficiently to embrace the combined effects of many varied and interacting factors, for instance type of processing machine, processing conditions, and cavity or die geometry, so as to satisfy criteria of acceptability that vary with the end-use requirements. Secondly, it tends to ignore the fact that an expansion of the enabling science would entail a proliferation of the structural measurements deemed to be necessary because, in that field as elsewhere, the progressive refinement of techniques uncovers new levels of influential detail that then demand deeper studies. Thus, the main outcome might be that the evaluation effort would tend to move away from the direct measurement of properties to the measurement of structure but with no reduction in the number of influential variables that would have to be taken into account and no guarantee, initially, that the trains of inference from structural data to consequential properties were reliable. Similarly, to explain structural details in terms of the processing variables would merely move the evaluation effort to the rheological and thermodynamic properties of the molten materials, which are as complicated as the properties of the solid state.

The issues were distorted during the 1980s and even more so in the 1990s by a recognition that fundamental studies and data generation activities were now so costly that they could not be supported at the levels that were appropriate during the high-growth-rate phase of the plastics industry. In view of the constraints on costs, intervention was necessary at various levels to develop more cost-effective strategies.

At the level of business strategy for instance, the primary producers are reluctant to generate properties data as liberally as hitherto and now tend to undertake exhaustive evaluations of property–structure relationships only when they have reached an understanding with a downstream operator in relation to a specific application. That strategy was adopted primarily to link evaluation costs directly to profitability but it also makes intellectual sense in that the property data arising from the totality of experimentation, testing and modelling activities may be regarded as constituting the outer surface of an ever-expanding sphere of knowledge which, for the utilizer, must invert into the inner surface of a sphere of knowledge of which only selected parts are required to collapse down to certain attributes critical for the application. A specific link between a primary producer and a downstream operator enables the evaluation strategy to be refined and focused and such operations benefit individual companies and, ultimately, those members of the general public who acquire the new or improved products. On the other hand, the results are usually confidential in the first instance and the data are unlikely to enter a database in the public domain at an early stage and therefore they do not benefit the industry in general or contribute to the enabling science in the manner and to the degree that was common in the 1960s. Furthermore, the majority of such collaborations are driven by the market aspirations of the downstream operator and are more likely to satisfy those particular commercial objectives than the wider ones of the primary producers with outlets in several industrial sectors. Therefore, the primary producers should always seek to incorporate the narrowly focused projects into the overall company strategy by judicious control of, and influence on, the experimentation; a clear distinction should always be maintained between properties and the property combinations that constitute attributes.

At the test/experimentation level there have also been innovations. The continual effort that had been expended during the late 1960s and early 1970s to generate comprehensive data to support the various downstream developments of thermoplastics artefacts was questioned on the grounds that a more comprehensive map of a property does not necessarily con-

tribute to bridging the gulf that lies between the properties of test specimens and the service performance of end-products. Since each test specimen has its particular flow geometry (which varies with the processing
conditions, the shape, the gating system, etc.), the status of any conventional
property datum relative to the service performance of an end-product is
questionable. Taken to the limit, that line of argument would lead to a policy
of direct testing of end-products but the data from end-product tests have
only local validity and therefore the practice is resorted to only when
special circumstances prevail. An alternative strategy is a compromise, the
use of test pieces of more complicated shape than the traditional ones but
simpler than most end-products, from which emerged the concept of 'critical basic shapes', see Supplement S1.11. In some respects that strategy
follows the long-established practices of the fibre-composites industry
which tests a hierarchy of structures, substructures, components, subcomponents, etc. and was also anticipated in the plastics industry in the late
1960s.[1]

There have also been developments at the level of databases and other
data displays, see Supplement S1.12, with moves towards standardized
formats for certain classes of data, but there are unresolved concerns about
the accuracy and precision of some of the data that are likely to be incorporated and also about the levels of detail for the data displays. Thus, for
instance, the high precision and fine detail that is now available in finite
element computations is out of balance with the limited accuracy of the
constitutive equations or other mathematical models by which the mechanical behaviour is summarized, with the accuracy and/or precision with which
the critical properties can be measured and with the levels of detail in which
they have probably been measured and stored. On the other hand, a truly
comprehensive set of data offering an extensive array of possible property
values covering the effects of the important influential variables may be
confusing rather than illuminating to a data utilizer who has no specialized
knowledge of the properties of plastics and only limited access to advice
from the data generator, but see also Supplement S1.12.

Reference

1 R. E. Hannah and J. A. Blanchette. 'Testing moulded and extruded parts.' *SPE Journal* **24**, (10), 102, 1968.

S1.11 Critical basic shapes

The cost of testing escalates with the increasing complexity of the data that are required as applications become more demanding and penalties for failure become more severe. The efficiency of data generation operations can be improved by three routes, i.e. abbreviated test procedures, the refurbishment of old tests and the development of new unconventional ones, in each of which the reduction of the time required for specimen preparation (the ultimate being the direct testing of mouldings with no intermediate machining operation) and the use of multipurpose specimens offer potential economies.

The argument for change has been supported by the emerging recognition that a modern test programme should accommodate the facts that a test specimen is a mechanical structure and the 'properties' derived therefrom are the properties of that structure (or that class of structure) and not the properties of the material *per se*. Thus, the classical concept of mechanical and physical property is questionable in relation to the generation of properties data for plastics and consequently the standard constraints on how such materials should be evaluated becomes less sacrosanct. For this class of material a property datum derived in one physical situation cannot necessarily then be utilized unconditionally in another merely by means of a mathematical adjustment equation; there has to be an independent validation of the transfer step.

The breakdown of what had been a traditional transfer process has ramifications for the choice of specimen geometries, the quality and precision of the tests, the subsequent processing of the data and their utilization thereafter. If a datum generated in a particular test has no general, wide-ranging validity as a material property datum, there may be little point in effort being expended on the attainment of high accuracy; equivalent effort expended on the generation of supplementary data, e.g. for other test configurations and other sample processing histories, may be a more effective alternative course that will provide a population of attainable property values which are separately appropriate for particular end-products and service situations. It follows, correspondingly, that parts of the classical strategies of experimentation and materials evaluation may be inappropriate in some instances and, in particular, the notion that specific specimen shapes guarantee high precision and accuracy has to be reassessed in the light of ceratin local circumstances.

Clearly, the flow geometry in the injection moulding of thermoplastic materials is a critical variable and it should therefore be treated as such in evaluation campaigns. In injection mouldings it is a function of the geometry of the cavity, the position and size of the gate, the runners and the processing conditions; in extrudates it is a function of the die geometry, the spider lines, the draw-down, etc. It varies also with the polymer architecture, the molecular weight, the molecular weight distribution and the composition of the plastic. All of those factors should be borne in mind during the design of any experiments and during the interpretation of results. Flow geometry features that commonly arise and are therefore of special interest include diverging flow, converging flow, flow along long, narrow channels, in thick sections, in thin sections, at abrupt changes of section, at abrupt changes of flow axis, parallel-flow knit-lines and opposing-flow knit-lines. Only a few of the flow geometry features will play a critical role in any one practical situation, but the combination rules and combination synergisms are as yet largely unknown.

The important flow features can be separately simulated in a set of simple mould cavities and, with careful design and some compromise, the mouldings can serve as test pieces which may be regarded as 'critical basic shapes'. They need not necessarily be radically different from the traditional specimens and it is probably better that they should not be. Thus, for example, the discs and other plaques that have been used either intact or as sources of conventional beams and bars in the past can be so gated as to incorporate knit-lines, to induce flow irregularities, different flow regimes etc. That was often done in the past as approximate simulations of practical end-product shapes but those earlier practices served specific downstream purposes and were largely isolated and uncoordinated activities.

Critical basic shapes offer the following advantages over conventional test pieces:

1 low-cost specimen preparation and easy test procedures, i.e. economies in the cost of evaluations;
2 discretionary incorporation of the major flow features and/or molecular alignments that affect end-product service performance;
3 data that have direct relevance to the performance of matched end-products.

On the other hand, there are certain penalties attendant upon an abandonment of 'ideal' specimen geometries, viz.

1 loss of precision and accuracy in individual measurements;
2 deviations from the strict definitions of properties;
3 possibilities for the proliferation of notional property values and data expressed in arbitrary units.

Ideally, critical basic shapes should simulate end-products or parts of end-products, but without sacrifice of generality and without the physical integrity of the results being compromised. Since they should be so designed as to incorporate the important flow features found in commercial mouldings, they will tend to yield property values that are different from, and usually inferior to, those yielded by conventional standard specimens (the usual reference data) and the rerating factors referred to above will then often be derating factors. However, the rationale is not straightforward because both the flow geometry and the shape geometry have to be simulated, which entails various approximations and introduces numerous sources of potential error. The strength of a knit-line, for instance, is reduced by long flow paths, high flow ratios, low melt temperatures, shape constraints on strain accommodation and inadequate packing pressure (magnitude and/or duration of influence). Cavity pressure itself depends on and varies with:

- rate of cooling and 'freeze-off'
- gate and runner design
- cleanliness and condition of the nozzle-valve and check-ring
- clamp setting and machine settings
- characteristics of the polymer melt.

See also Chapter 2, Supplements S2.5 and S2.7.

Another factor that complicates the link between the properties of a conventional test specimen or a critical basic shape and those of a service item is the size effect, see Chapter 8; if a critical basic shape is substantially smaller than the item that it supposedly simulates, it may be tougher, because the propensity for a crack to develop depends on the ratio of the stored elastic energy (or, more correctly, viscoelastic energy), which is proportional to the cube of the linear dimensions, to the energy demand for the creation of fracture surfaces, which is proportional to the square of the linear dimensions. An alternative qualitative argument develops as follows.

The size of the plastic zone that usually develops at a crack tip is, to a first approximation, a material property and independent of the size of the cracked item. If the size of that zone is similar to a lateral dimension of the item the crack is unlikely to develop as a brittle fracture; it follows that the larger the item, the greater the probability of brittleness.

The current position is that critical basic shapes are often used casually as simulations of particular end-products but increasingly in a more structured manner which works towards quantitative validation of the method, see reference 1 for instance. Correlations between flow geometry and the stiffness and strength anisotropies that are found in end-products are now fairly well established; flow irregularities, which are the instability features of flow geometry, relate to toughness and have not yet been documented so systematically though certain interesting correlations have been recorded for impact resistance, see Chapter 11. Certain issues remain unresolved; is there, for instance, a definable set of basic flow geometries from which all practical mouldings could be synthesized, could a set of combination rules be devised to rationalize and quantify that synthesis and could interactive links be established between product designers and materials evaluators to match critical basis shapes to commercially practical structures?

Once comprehensive information on the weakening effect of flow irregularities has been assembled, the only issue remaining unresolved would be the extension of the flow geometry and flow irregularity concepts from essentially two-dimensional plates to three-dimensional items. In the past, and for many years, elementary versions of the so-called 'falling weight' impact test were used widely for the testing of moulded articles, as a means of quality control and quality assurance, and several primary producers developed in-house tests of simple moulded shapes, such as open-topped boxes. The stiffness of such an item, its impact energy and the character of the failure depends on where it is hit and how it is supported so that the tests are arbitrary but nevertheless the results were claimed to correlate with the inherent toughness of the material from which it was made and also with the toughness of other moulded shapes. Some such correlations have been published but there seems to have been no systematic attempt to develop the method quantitatively. It should be noted, however, that even though the flow geometry and flow irregularity aspects of three-dimensional mouldings had not been resolved analytically or even placed on a sound pragmatic base the severely practical and technically limited tests that have been used nevertheless sufficed to support the plastics moulding industry during its high growth phase and to ensure that the end-products were fit for their purpose.

In some respects a testing strategy for thermoplastics that incorporates the use of critical basic shapes is similar to a long-standing one for fibre composites that uses 'subcomponents'. It differs mainly because the issue of flow geometry variations in thermoplastics is less tractable mathematically and less amenable to control than the corresponding issue of fibre alignments and structural assembly in fibre composites.

Reference

A. S. D. Pouzada and M. J. Stevens. 'A case study on the prediction from subcomponents of the flexural behaviour of commercial mouldings.' *Plastics and Rubber: Processing and Applications* **6**, (3), 209, 1986.

S1.12 Data generators, data utilizers and information pathways

The plastics testing community can be divided into data generators and data utilizers, but with the proviso that generators utilize from time to time whilst utilizers are often obliged to generate when the data that they receive are insufficient or inadequate for their purpose. The generators are plastics producers, contract testers and academic researchers, whereas the data utilizers are designers, converters and the users of plastics end-products. The two groups are connected by various information pathways, some of which are of dubious efficiency. Furthermore, the various classes of data generator tend to have different motivations and priorities. Thus, for instance, testing carried out in academic institutions is often part of a postgraduate course of study, in which the attainment of research skills is of greater importance than test method development or general data archives, so that many interesting and potentially useful techniques are never developed beyond the point at which a higher degree is awarded. Similarly, contract testing is constrained by the scope of the remit and the funding and any improvements to a test method or a procedure that arise under the contract may never enter the public domain.

In industry, data generation activities are constrained nowadays by the cost and the utilizers often claim that the data that they receive are inadequate in scope and bear little relationship to the service performance of end-products. That complaint is a legitimate one in that most of the property values quoted in public domain data sheets, etc. are still likely to be approximations to the upper bounds of the possible values, by virtue of the test methods and specimen geometries used in their derivation, whereas the service performance of plastics end-products is often governed by the lower bounds. Even so, criticism of the primary producers is largely unfair because they originally invested large resources on testing and research, established the basic facts about property variations and published the results of their investigations even when the logical consequence of that was an increased demand for test data. Also, certain primary producers have collaborated to produce a common database[1,2] which is of great use to the utilizers (and presumably profitable to the creators!) and more expansive databases could follow that initiative.

Several factors contribute to the disparities between intention and achievement in data generation. One is the complexity of the properties; their great sensitivity to the processing route and conditions has created a demand for ever more data that cannot be fully satisfied by traditional methods within acceptable cost limits. A second factor is the different and conflicting vested interests of each class of generator and utilizer, which influences its operational strategies, tactics and expectations. Additionally, within each class competition between the various parties affects courses of action and levels of collaboration on matters such as standardization. The data generators are mainly concerned with the properties of materials whereas the utilizers tend to be concerned with the attributes of service items. A third factor is a burgeoning requirement for enhanced data to cover increasingly exacting downstream applications that entail the materials being used at the limits of their performance capability where unanticipated deficiencies may be manifest. Considering all the circum-stances, it is possible that some generated data may not be acceptable to a recipient database, particularly if the 'properties' data generated are not matched to the 'attributes' required by the utilizers. However, specific arrangements between primary producer and downstream operator, as referred to in Supplement S1.9, can avoid mismatches and partly circumvent the cost constraints by directing the data generation activities toward particular components and materials. Only strictly essential data are then generated so that the information pathway carries prompt feedback and the data demand evolves directionally. Relatively recent successful outcomes of such collaboration include large components in domestic equipment and automobiles (e.g. washing machine tubs and automobile bumpers).

However, in even the simplest case of a pathway between a collaborat-ing generator and utilizer distinctions have to be drawn and utilizer choices made between actual data and processed data (extrapolated, interpolated, inferred, idealized and so forth). A mechanical property P is a function to time, temperature, molecular state of the sample (or the commercial end-product) and is likely to feature as a multivalued function $\{P\}_{store}$ in any particular database and consist of a measured component $\{P\}_{meas}$ and an inferred component $\{P\}_{inf}$, the elements of which will be variously reliable and accurate. The extraction of the relevant bits of information on possibly several such properties for a downstream application is largely a matter needing skilled human judgement to qualify the information for which an intellectually isolated electronic transfer of data from generator to utilizer is a poor substitute. Irrespective of the mechanisms of data generation and data transfer, clarification of the status of all archived and transmitted data should be obligatory It is not so at present; certain accreditation bodies are moving towards stipulations that should establish good practices with regard to record keeping and before that is fully achieved it behoves all data generators to adopt best practice in that respect. Simultaneously they should develop utilizer-friendly formats,[3,4] and seek also to merge new data with the voluminous archive data even though they are likely to be more reliable than the latter because of being a later product of the learning curve.

References

1 K. Oberbach. 'Fundamental data tables and databases (®CAMPUS) – a challenge and opportunity.' *Kunststoffe German Plastics*, 1989/8.

2 H. Breuer, G. Dupp, J. Schmitz and R. Tüllmann. 'A standard materials data bank – an idea now adopted.' *Kunststoffe German Plastics*, **80**, (11), 1990.

3 S. Turner and G. Dean. 'Criteria in determining mechanical properties data for design.' *Plastics and Rubber: Processing and Applications*, **14**, (3), 137, 1990.

4 P. J. Hogg and S. Turner. 'Critical issues relating to the quality of properties data.' p.58 Technical papers, AMMA Third International Conference, 'Materials innovation and their applications in the transportation industry ATA-MAT 91.'

S1.15 The development of standard test methods

By the middle of the nineteenth century, the testing of materials had acquired an aura of prestige and a reputation that implied decisive trial, critical examination and thorough characterization. That reputation and the commercial interest in 'fitness for purpose' ultimately entailed the standardization of the methods of test. The Standards Institutions now exert an important influence both on laymen, who tend to view them with favour as guardians of their standards of living, and on some scientists, who see them as a basic framework against which their own experimental techniques may be devised and compared. However, the route by which a standard test method is developed is not usually as scientifically respectable as many people might wish or as most people believe. A typical development path is as follows.

An individual or a group with a common interest, e.g. coworkers in a particular company, develop a test routine that serves, or seems to serve, a local technical or commercial interest and other groups accept it, reject it or impose a compromise as a matter of commercial self-interest, after which a draft standard is prepared and supporting evidence is provided. The method may be arbitrary, misleading and results may be precise rather than accurate; it must be as simple and as inexpensive to conduct as possible and 'non-threatening' to a wide range of interests. Despite those limitations, the standards are helpful rather than otherwise and whilst there are obvious technical reasons why particular test practices should deviate from the Standards those deviations should always be justifiable, should always be identifiable via qualifying information and should retain as many of the critical features as possible (particularly temperature, testing rate, dimensions and maximum strain).

A common source of almost inadvertent non-compliance is the molecular state of the sample, to which reference has been made at various points in the preceeding sections. The sensitivity of the influential structural features to the processing conditions is often accommodated within standard procedures for tests on particular materials by stipulations concerning conditioning periods, annealing treaments and even direct recommendations about the processing conditions. Such stipulations have the merit of reducing variability in test results and they are often a reasonable compromise between what would be ideal and what is practicable, but they can sometimes inadvertently consolidate an outdated practice that should really be replaced.

In general, standardized test practices are followed by industrial scientists provided that the resulting data are in harmony with the in-house archive data and with the overall vested interests of the organization. Academic scientists are less compliant, partly for cultural reasons, partly because their work may be unique and sometimes merely from expediency. The main subsequent difficulty rests on how results obtained in non-standard procedures may be utilized safely in a wider context where they may be at variance with corresponding data from a standard test. Ideally, the issue should be resolved by open debate but there are many practical obstacles to such cooperative and collective activity.

A recent development has been Standards or draft Standards for the presentation of data, see also Supplement S1.12. The need for this has arisen because plastics' producers have tended to use different formats and different scopes for the in-house technical data that they generate in support of their products, and data utilizers then cannot readily compare data sets from different sources.

Those very important developments progress only slowly because of the nature of international moves towards consensus but, at a more local level agreements on presentation formats have been reached relatively rapidly and to the mutual advantge of the participants as, for instance, when several companies within one national boundary identified a common objective. The success of one such group is mentioned in Supplement S1.12; the general approach was along the lines described in references 1 and 2 of that Supplement and essentially provided a template for subsequent ISO activity on data presentation formats. That development sequence is, of course, the route by which testing standards have always emerged.

The emerging standardization of data presentation formats with supporting items such as sample description and data quality is not now sufficient. Electronic transfer of data between existing databases, comparisons between different classes of material and compatibility, etc. are under sctutiny for standardization,[1] driven originally by commerce and engineering considerations outside the plastics industries but impinging now on the plastics testing community and not necessarily advantageously if any of the important differences between the characteristic properties of plastics and those of other materials are diffused in the interests of a common data format.

Reference

1 V. Wigotsky. 'The road to standardization testing.' *Plastics Engineering*, 22, April 1995.

General comments on modulus, ductility, stiffness and toughness

2.1

In Chapter 1 it was stated, without specific details being given, that plastics are non-linear viscoelastic in nature and that end-products made from them are often anisotropic. Thermosetting plastics and fibre composites are also viscoelastic but generally to a lesser degree owing to the cross-links in the former and the fibres in the latter, though variable via cross-link density, molecular architecture and fibre alignments. As a consequence of those characteristics, and as will be discussed in later chapters, the assessment of the mechanical properties of both plastics materials and end-products manufactured from them is a complicated task. A property such as modulus, for instance, may be defined and measured in several ways and have a wide range of values depending on these circumstances. Thus, a single datum quoted as a property value is likely to be an incomplete and possibly biased indicator of that property.

Despite that limitation on the validity, accuracy and precision of a measured property datum, even when the test method is a reputable, standardized one, there have been innumerable occasions on which assessments and decisions have been based on such curtailed data, and one can assume that serious errors of judgement have been rare. The reason is probably that, nowadays at least, single point data are viewed and judged against a background of archive data and case histories, and with due caution when the application is novel or exacting.

2.2

Modulus and ductility are two amongst several multivalued properties that are important in the context of downstream service. Setting those complications aside and using terms such as 'modulus' in a general sense, plastics have very much lower moduli and strengths than most metals, see Chapter 3 and Table 3.8.1, though they are sometimes tougher. Since load-bearing is often either a primary function or a secondary function of a plastics end-product, even small modulus enhancements can be commercially advantageous and therefore 'modulus' is an important consideration in activities as diverse as materials comparison, materials selection, materials product development and end-product design. The issues are not necessarily resolvable in a straightforward manner, first because an appropriate property value has to be chosen from the many possible values and secondly because an adequately stiff end-product or component may be more readily and economically attained through the agency of shape or section thickness than through a high modulus. Cost per unit volume and the feasibility of the melt-processing stage then have to be included in the deliberations.

2.3

It often transpires, particularly with thermoplastics, that increased modulus is accompanied by reduced ductility, which is potentially disadvantageous and requires a proper balance to be maintained between the two properties and between the stiffness and the toughness that they respectively engender in an end-product. There are exceptions to the reciprocal relationship because a modulus measurement produces a weighted mean value over the body of the specimen whereas a ductility measurement may be biased by a local failure event but the general rule is a consequence of the molecular architecture. The natural vibrations of the various species of molecular segments are biased by an applied stress and the rate of subsequent equilibration depends on the segmental mobilities which, in turn, depend on the architecture; a low mobility is manifest as a relatively high modulus because the strain response develops only slowly but by the same mechanism the ductility may be correspondingly low because molecules may break before they can attain new equilibrium positions. A continuum mechanics explanation that emerges from the theory of fracture mechanics is as follows: modulus and yield strength increase in the same sense as the polymer architecture, the temperature and the rate of stressing are changed, since the yielding process is merely a final consequence of increasing deformation (unless brittle failure intervenes), but as yield strength increases the probability of failures being brittle increases because the size of the plastic zone that forms at the crack tip and which affects the propensity of the crack to grow is inversely proportional to the square of the yield strength, see Chapter 8.

The loosely reciprocal relationship between modulus and ductility was recognized long ago and various relatively crude tests have been deployed to quantify the matter. Results typical of a past era, in the form of Young's modulus in flexure vs. notched Izod impact strength for a range of polypropylenes are shown in Fig. 2.3/1. It was published by a plastics

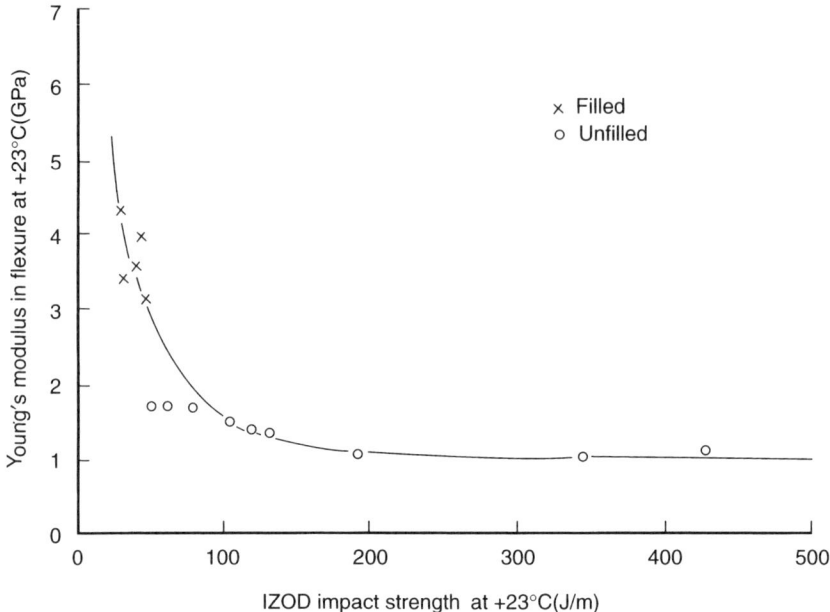

2.3/1. The relationship between modulus and ductility for various polypropylenes. The data, taken from a manufacturer's published information, shows the relationship between Young's modulus in flexure and Izod impact strength at room temperature.

producer many years ago and is a crude indicator of the modulus–ductility relationship that can be obtained quickly and at low cost. The relationship shown here would have been similar in shape but different in detail had other measures of modulus and ductility been used, as is likely in modern evaluation programmes, but the basic feature of increased stiffness being accompanied by decreased toughness persists. Figure 2.3/1 and the many similar sets of data generated in the past are of questionable accuracy because the tests used were chosen for their simplicity and ease of operation rather than for their precision and physical significance. They are also now known to have been less definitive than was thought hitherto because, with rare exceptions, they inadvertently disregarded the distorting effects of variations in the flow geometry by having the modulus and the ductility measured on different types of test specimen and sample. On the other hand, they still provide useful insight into the likely service performance of plastics materials, especially when they can be considered in relation to archive data and practical experience of service performance.

Also, the modulus vs. ductility relationship in Fig. 2.3/1 is only an approximate description of reality in that it disregards processibility, which varies with the class of polymer and from grade to grade within each class, which is liable to affect modulus and ductility to different degrees and which is usually dictated by unrelated commercial considerations. In thermoplastics the inherent toughness benefit of a higher molecular weight can be completely offset by detrimental effects arising from the higher melt viscosity, e.g. internal strains, flow irregularities and anisotropy that may be engendered in a moulding or extrudate. Thus, though Fig. 2.3/1 or some analogous diagram can be taken as an elementary prototype of the linked system of mechanical tests depicted in Fig. 1.4/1, Chapter 1 or even as an abbreviated and economic version of that system, an absent third axis is a deficiency, albeit one that is nevertheless partly compensated for if both types of test are conducted on a service item or a section cut from one and if the coordinates are stiffness and toughness rather than modulus and ductility. See Supplements S2.5 and S2.7.

In thermosets, cross-link density plays a role analogous to that of molecular weight in thermoplastics. In fibre composites, volume fraction and spatial distribution of the fibres, interfacial adhesion and general consolidation affect both the stiffness and the toughness of an item, but there is no obviously reciprocal relationship between the quantities and the toughness has no support from ductility which is essentially meaningless in a fibre-composite context.

2.4

The 'modulus' and the ductility can be explored in whatever detail is deemed to be necessary, but abbreviated programmes are usually adopted because comprehensive evaluation is expensive and because stiffness and toughness are only two of several sometimes conflicting performance requirements that may be specified for an end-product, for example low cost, high strength, low thermal expansion coefficient, good dimensional stability, high creep resistance, high solvent resistance, low flammability.

Even so, in the light of the general character of the mechanical properties of plastics, a minimum strategy for modulus/ductility evaluation should be one that assesses how well the ductility is retained at lower temperatures and how well the modulus is retained at higher temperatures. Other issues of similar importance are the rate at which 'apparent modulus' decreases with increasing elapsed time under an applied constant excitation (i.e. the propensity to creep or to relax), the decrease in toughness under sustained or intermittent stress (i.e. the susceptibility to creep rupture and fatigue failure), see later chapters, and the embrittling effect of stress triaxiality. An economic modulus/ductility evaluation strategy that extends beyond the confines of Fig. 2.3/1 is outlined in Supplement S2.4.

2.5

Indications of such trends can be obtained from arbitrary but standardized procedures such as a heat distortion test (HDT) and a low temperature brittleness test, both of which have been widely used in the past, though the data derived by such methods must be viewed with caution because the procedures are arbitrary and flawed. For instance, the deformation of the specimen in the heat distortion test is a compound response to elapsed time under stress and simultaneous rise in temperature, which cannot be separately quantified, whilst the results from the low temperature brittleness test depend strongly on the quality of the specimens and usually relate only to items wetted by a liquid coolant.

Table 2.5/1 summarizes the overall effects of test or service variables, composition and sample state on the modulus and the ductility of thermoplastics, in so far as such judgements can be based on results from particular, conventional test specimens. The second and third sections of the table list those factors that can be introduced by physical or chemical modification of a polymer. Only one, higher molecular weight, offers any possibility for simultaneous enhancement of both modulus and ductility and that route has a largely unpredictable outcome because of the multifaceted influence of molecular weight and molecular weight distribution via the melt processing stage, but see also a footnote to Table 2.5/1.

Table 2.5/1 Balance between modulus and ductility in thermoplastics

Operative factor	Effect on modulus	Effect on ductility
Reduced temperature	Increase	Decrease
Increased straining rate	Increase	Decrease
Multiaxial stress field	Increase	Decrease
Incorporation of plasticizer	Decrease	Increase
Incorporation of rubbery phase	Decrease	Increase
Incorporation of glass fibres	Increase	Decrease[b]
Incorporation of particulate filler	Increase	Decrease[b]
Increase of molecular weight	Little direct effect	Increase
Increase of crystallinity[c]	Increase	Decrease
Copolymerization	Decrease[a]	Increase[a]

[a] Copolymerization usually results in this change. Where the comonomer is a stiff unit, the effects may be reversed, but modulus and ductility still change in opposite senses.
[b] Phase separation can produce a prefailure mode of pseudoplasticity, sometimes characterized by higher energies to failure than that of the unmodified matrix.
[c] Where relevant.

The modulus, ductility and melt viscosity of a material translate into the stiffness, toughness and ease of production of an end-product made from it. A comparable inverse relationship between stiffness and toughness can often be discerned but the relationship between the three attributes is more complex than the properties set out in Table 2.5/1. There are three considerations to be made, first because skilful choice of shape can sometimes confer stiffness without commensurate sacrifice of toughness; second, because flow irregularities in a moulding or extrudate can reduce the toughness locally without increasing the stiffness; third, because crystallinity and molecular orientation vary in complex ways with the processing conditions and the flow geometry even to the point where both modulus and ductility are affected in the same sense, see also Supplements S2.5 and S2.7.

2.6

Variations of established test procedures have been developed over a long period with the result that an almost bewildering range of tests can now be deployed in an evaluation. This was mainly driven by newly developed plastics and by old plastics in new applications. Many of those tests were introduced as *ad hoc* solutions to particular problems rather than as unifying methods of testing, and there is now some overlap in their respective roles. On the other hand, the increased range of test procedures provides options for the use of other datum pairs, for example, yield strength and critical stress field intensity factor (see Chapter 8).

It can be argued now that the most pressing need is for the use of common specimen geometries for as many of the datum pair tests as is practicable. At a single step this would remove processing history and flow geometry as sources of irregularity in a modulus–ductility relationship and it would also simplify and cheapen the provision of test pieces which has proved to be a major burden for plastics producers acting as data generators. However, such a step is not entirely straightforward because it raises various associated issues, e.g. the need to modify certain currently popular test practices, the possible development of novel test configurations and the systematic use of 'critical basic shapes' (see Supplement S1.11). Some modified tests that offer datum-pair possibilities for commercial shapes, e.g. pipes, are discussed in later chapters.

2.7

In the simplest cases, the common specimen geometries referred to in Section 2.6 are conventional test pieces, e.g. bars, beams and plates, and possibly specimens incorporating a knit-line at some critical point with respect to the test configuration. Later, the common specimen geometries could be less conventional, i.e. the so-called critical basic shapes referred to in Section 1.11 and Supplement S1.11.

Ideally, the stiffness–toughness balance should be assessed by modulus and ductility measurements on various appropriately shaped mouldings, see Supplement S2.7. The coordinates of the points plotted in diagrams such as Fig. 2.3/1 are then found to vary with the shape of the test piece and with the processing conditions generated by the flow geometry. Thus, an order of merit derived for one particular flow geometry may not be retained with a different flow geometry.

2.8

The use of critical basic shapes instead of standard specimens introduces the concept of flow geometry as a critical variable in evaluation programmes for thermoplastics. Additionally, it changes the emphasis from the supposed properties of materials to the properties of simple structures. However, the properties of one structure cannot necessarily be translated into the properties of a different structure; the standard specimens are themselves structures and the properties derived therefrom do not relate directly to the material (Sections 1.7 to 1.11).

2.9

The critical basic shapes rationale is similar for fibre composites, with the volume fraction and spatial distribution of the fibres replacing flow geometry as a critical variable. The composites industry is far ahead of the thermoplastics industry in its appreciation of this point and uses defined specimen geometries and fibre alignments as critical basic shapes without using that specific label. The test pieces are often regarded as substructures. However, that awareness has not enabled the composites testing community to avoid a proliferation of some classes of test and also a neglect of some areas.

2.10

This chapter and its Supplements discuss only modulus in relation to ductility, which is a fundamental issue. There are, however, many other datum pairs that are important as indicators of the commercial acceptability of particular products, e.g. transparency with toughness, surface texture with toughness, flexibility without taint and so on. Each particular case requires its own datum pair as a minimum criterion and, judging from the many examples described in the literature, most of the contributory tests are arbitrary, just as the original tests for the modulus–ductility datum pairs were.

It would be unhelpful, and ultimately futile, for those pragmatic down-stream test procedures to be criticized on fundamental grounds – it can only be hoped that some effort be directed towards linking the test data, which refer usually to attributes, to the properties derived from authenticated procedures, or at least quantifying them in a wider context than a single dedicated use.

S2.4   A modulus–ductility evaluation strategy for thermoplastics

Throughout the period of rapid growth in the plastics industry (i.e. 1950–1970) various evaluation strategies were developed but tended to become progressively more elaborate as the industry grew and as plastics were increasingly used in exacting applications. Cost considerations also became a constraint. A constantly recurring theme in the development of new or improved materials has been the attainment of an adequate balance between the modulus and the ductility of a material, or between the associated stiffness and the toughness of moulded end-products, in view of the reciprocal relationship between the two properties.

Initially, it was commonly believed that the so-called 'dynamic modulus', nowadays more correctly referred to as the complex modulus, see Chapter 5, could give insight into both the modulus and the ductility aspects of the mechanical behaviour since they are both governed by the mobility of the molecules. However, whereas modulus is either independent of molecular weight or indirectly dependent on it, ductility is very sensitive to it. This applies up to some limiting point at which the packing efficiency and the molecular chain mobility are progressively impaired as the molecules get longer. The impaired mobility is also manifest as a higher melt viscosity, with the penalties of processing difficulties and unfavourable flow geometries. Thus, it was finally conceded that complex modulus alone could not adequately quantify the modulus–ductility characteristics of a sample.

It subsequently became apparent also that elementary modulus data should be supplemented by creep or relaxation data and that elementary strength data should be supplemented by creep rupture and fatigue data, etc. Stress relaxation and creep signify a reduction in 'modulus' with increasing time under stress that is not necessarily accompanied by an increase in ductility and similarly creep rupture and fatigue failure may signify a reduction in ductility that is not accompanied by an increase in modulus. The comprehensive testing programmes that were consequently considered to be essential were an expensive burden for the data generators (mainly the primary material producers) and, in periods of low growth, consolidation and restructuring of the plastics industry, some retrenchment has occurred in response to the need to reduce operating costs. Thus, the modern abbreviated programmes have reverted in part to the cruder tests of an earlier era, though with certain precautionary refinements and an awareness of the great influence that sample state and flow geometry can exert on the outcome of tests.

Setting considerations of sample state and flow geometry aside, the purely mechanical aspects of a minimum cost strategy for the evaluation of the important properties, modulus and ductility, are now:

1 Initial measurement of modulus and ductility by standard, arbitrary tests to provide a datum pair for comparison with other materials. This is often a commercial rather than a technical necessity.
2 Additional tests to clarify:
   (a) the effect of raised temperature, higher stress, sustained stressing on 'apparent modulus';
   (b) the effect of lowered temperature, high stressing rates, intermittent stressing and stress triaxiality on the ductility.

If sample state and flow geometry are to be considered, as they should be if the evaluation is to relate to expectations for downstream service, the number of tests can escalate to an unacceptable degree unless deliberate steps are taken to limit the options of specimens chosen for their close correspondence to service items. At a superficial level that is not a new idea; particular mouldings have been commonly used to provide data relating to commercial processing practices and the service performance of end-products, but they seem to have been used only in limited contexts. What is now feasible is the use of such mouldings to build up a quantitative and comprehensive picture of behaviour in various classes and subclasses of end-product and application (see Supplements S2.5 and S2.7 for more details). In the restricted context of a minimum cost strategy, the best practice is for the modulus and the ductility tests to be carried out on the actual service component that is of particular interest, on a critical part of it or on a simulation, i.e. for the stiffness/toughness relationship rather than the modulus/ductility relationship to be investigated. Item 2 of the minimum cost evaluation strategy set out above should therefore be extended by:

(c)   the effect of relevant, practical flow geometries on both the modulus and the ductility.

The precision and the accuracy of the supplementary data should be at whatever level is appropriate in relation to the purpose of the evaluation but should be known to the investigator and stated with test results. The precision of an arbitrary test can be high and can often be easily maintained at a high level by strict compliance with an established procedure; the accuracy, on the other hand, is often highly questionable and difficult to establish and to quantify.

S2.5 The balance between stiffness and toughness in injection-moulded end-products

The stiffness and toughness of a thermoplastics end-product are governed by the state of molecular order, e.g. the degree of molecular alignment and the incidence of flow irregularities that prevails in it. Those features, in turn, depend on the changing condition of the molten and solidifying plastic during the processing stage. The primary influential factors are the viscoelastic and thermal characteristics of the molten and solidifying plastic mass in the cavity. However, the practical factors are the processing conditions (cavity pressure, melt temperature, cavity temperature and cooling rate), the geometry of the cavity and the geometry and position of the gate or gates.

The many influential factors are interactive and therefore the simple correlations listed in Table 2.5/1 tend to be obscured or distorted in service situations. For example, on kinetic grounds the faster the cooling rate through the solidification stage the lower the degree of molecular order and therefore the tougher the moulding, but rapid cooling (via heat transfer to a relatively cold cavity surface) tends to encourage molecular orientation in the surface layers and may also cause premature 'freeze-off' with consequential reduction in cavity pressure, melt starvation in remote regions and internal strains, all of which are detrimental to toughness. Mouldings with varying section entail correspondingly differing cooling rates (unless the layout of the cooling channels has been designed to compensate for that) and consequential point-to-point property variations which further complicates the stiffness/toughness relationship.

There are several other practical situations in which the stiffness/toughness outcome is not straightforwardly predictable. Long, narrow flow paths encourage uniaxial molecular orientation which confers enhanced toughness and possibly enhanced stiffness along the axis of orientation but reduced toughness and possibly reduced stiffness in the tranverse direction, see also Supplement S2.7 and Chapter 7. Balanced planar orientation, which can arise in some plate-like mouldings, gives uniform enhanced toughness and possibly enhanced stiffness in the plane but reduced stiffness in flexure (due to shear deformations) and some tendency for failure by delamination. However, those orientation effects are either diluted or radically modified in thick-walled mouldings because the pattern of molecular alignment varies through the thickness.

Knit-lines are similarly problematical in that, whilst they are usually regarded as flow discontinuities and therefore points of local weakness or even brittleness, they do not necessarily behave in that way; weak knit-lines arise in several circumstances, namely

- when fusion of the two melt fronts is incomplete owing to the melt temperature and/or the pressure being too low, which occurs if the flow ratio is large;
- when melt flow occurs (usually in the core region) in the direction transverse to the main flow axis;
- when the venting of displaced air is inefficient, which causes scorching due to adiabatic heating of the entrapped air and/or incomplete filling of the cavity at the knit-line (generally notch-like surface defects);
- when the thermal conditions in the knit-line region lead to an atypical molecular state of order locally;
- when the venting mechanism allows a thin flash-line containing highly oriented molecules to form.

See references 1 and 2, for instance.

When such defects are eliminated, knit-lines are not noticeably weaker or less ductile than unblemished material in a similar state of molecular disorder, but the elimination of all the potential sites of weakness in a large moulding is virtually impossible and a satisfactory moulding is often largely dependent on designers arranging that the knit-lines occur at points where the anticipated service stresses are likely to be small and inadvertent stressing unlikely to occur. The main processing defects in thermosets relate to entrained voids and incomplete cure during the cross-linking phase, both of which can be minimized by appropriate mould design. In fibre-composites, unwetted fibres, misplaced fibres and poor consolidation are defects additional to those inherent in the matrix.

Those numerous and interacting factors pose difficulties for an investigator designing an experimental programme, first because its range may possibly need to be more extensive than the objective of the tests ostensibly requires and second because results arising from an arbitrarily resticted programme may be atypical of the behaviour as perceived by the plastics community at large.

References

1 R. Boukhili and R. Gauvin. 'Investigation of weld-line fracture of plastics injection mouldings'. *Plastics, Rubber and Composites: Processing and Applications*, 11, 1, 17, 1989.
2 S. Piccarolo, F. Scargiali, G. Crippa and G. Titomanlio. 'Relation between packing flow and mechanical properties of weld zones in injection-moulded thermoplastic polymers.' *Plastics, Rubber and Composites: Processing and Applications*, 19, 4, 12, 1993.

S2.7 Stiffness/toughness – critical basic shapes

The overall effectiveness of the procedures that are used for the assessment of the balance between modulus and ductility is impaired by the use of different specimen geometries and flow geometries in the two types of test. That deficiency in the experimental procedure can be rectified in some instances by the use of a common specimen geometry for the modulus and the ductility tests. The usefulness of the derived datum pair is enhanced further if that common specimen is a 'critical basic shape', see Supplement S1.11, or a piece cut from one, i.e. if the moulding contains some feature that is commonly found in end-products. Flat, apparently featureless plate-like mouldings may enjoy the status of critical basic shape if they contain important flow geometry features that arise from the feed system and the gate geometry; they include the following:

- single-feed weir-gated plaque;
- double-feed weir-gated plaque;
- coathanger-feed weir-gated plaque;
- single-feed edge-gated disc;
- double-feed edge-gated disc (two knit-line types);
- double end-gated tensile bar;
- picture frame with various thickness and gating options.

Each of those may have surface features such as ribs, grooves and bosses, and have internal inhomogeneities such as foamed cores and knit-lines, all of which can confer unusual distortions of the stiffness/toughness balance and confound the apparent simplicity of Table 2.5/1.

Such mouldings, apart from the picture frame, may be tested in flexure as entities, firstly at low strain for a plate stiffness datum and secondly by impact for a ductility datum as discussed in later sections. The same procedure, using one specimen successively to generate a modulus datum and then a ductility datum, can be adopted for test pieces cut from larger items, to determine the variation of properties and property balances from point to point in a moulding. The picture frame moulding is a useful source of dual purpose specimens that enable the effects of flow ratio on the modulus and the ductility to be assessed, see reference 1, for example.

In general, the use of a common specimen geometry and/or a common flow geometry for the two tests gives information about the material in a particular state of molecular order. The comparable assessment for a moulded end-product may be more complicated because the critical region governing the stiffness of the product may not be the critical region for the toughness and the two regions may have radically different flow geometries. For example, the propylene–ethylene copolymers used in moulded crates for beer bottles have to have a high resistance to creep (high stacks of full crates, possibly high temperatures) and a high resistance to impact (casual impacts during off-loading on cold winter days), but the creep resistance relies largely on the long-term stiffness of the vertical struts at each corner of the crate whereas impact failures arise mainly in the side panels, at corners and at other stress concentrators where the flow regimes are very different from that in the struts. In that particular case, one common flow geometry for the two classes of test would not suffice; axial compression on end-gated cylinders and flexed plate impact tests on variously gated plates would provide useful datum pairs that might suffice to discriminate between competing and closely similar materials, but the adequacy of the transverse toughness of the cylinders (or corner struts) and the stiffness of the plates (or panels) should also be ascertained, i.e. two sets of datum pairs and a cross-correlation.

The use of datum pairs for simple shapes, e.g. flat plates, should be straightforward though the individual points in Fig. 2.3/1 would be replaced by a loose cluster of points reflecting the variations in the datum pairs attributable to different flow geometries. When the shape is not simple, e.g. a ribbed plate, the reciprocal relationship of stiffness to toughness may be distorted. The stiffness should follow the elementary laws of mechanics but the toughness does not conform to a well-defined pattern because the local molecular orientation and anisotropy that arise as a result of the flow geometry or flow irregularities generated by a moulded feature may either accentuate any related geometrical stress concentration or mitigate it, depending on the local circumstances.

Those opposing possibilities complicate the analysis and also explain why the service performance of an injection-moulded article may differ from expectations based on the data generated in laboratory tests on conventional, standard specimens. However, practical experience supported by admittedly imperfect experimentation over many years led ultimately to the emergence of various design principles for gates. Thus gates for commercial mouldings should:

1   give even filling across the cavity; that can rarely be achieved and a more modest target is uniform filling across critical areas;
2   enter at, or close to, the thickest section;
3   be large enough to ensure fast filling of the cavity, to enable it to be packed before freeze-off occurs (high injection pressure is needed to enable a fast filling rate to be attained but the pressure must be reduced, usually to between 50% and 75% of the peak, for the packing stage).

With reference to the third point, gate size is a compromise – small tunnel or pin gates leave inconspicuous scars but may freeze off prematurely whereas larger sprue gates leave a noticeable scar unless an additional finishing operation is carried out.

There are corresponding design rules for commercial cavities and ideally those rules should not be violated in cavities for test specimens except when specific features are being investigated. Thus, for example,

1   flow fronts should not become stationary at any point in the cavity during the filling phase;
2   the flow into a reinforcing rib should be along its axis;
3   high flow ratios should be avoided;
4   transition radii at changes of section should be substantially less than the thickest section (e.g. one-half or less).

In view of the many factors affecting the quality of injection mouldings and end-products produced by other routes, and consequently the mechanical properties manifest by them, it is hardly surprising that the properties measured on traditional test pieces often differ from those of the end-products. An investigator should always bear in mind that a property value measured on a test piece relates to that test piece but not necessarily to any other item. In the absence of direct testing, the properties of end-products have to be estimated via trains of inference. Critical basic shapes are one stage closer to end-products than conventional test pieces and the connecting trains of inference may therefore be more reliable.

Reference

1   R. A. Chivers, D. R. Moore and P. E. Morton. 'The influence of flow path length on the stiffness and toughness of some engineering plastics.' *Plastics, Rubber and Composites: Processing and Applications*, 15, 3, 145, 1991.

3

Modulus and stiffness: general principles

3.1

'Modulus' relates to a material and translates into the stiffness of an article made from that material when the dimensions and load configuration are taken into account. For some classes of material, modulus can be regarded as a physical property the value of which should therefore be independent of the dimensions of the test specimen. However, this is not usually so for plastics because the properties vary with the degree of molecular order and other structural factors which, in turn, depend on the processing conditions and the flow geometry, see Chapter 1, Sections 1.5 and 1.7. Consequently, the stiffness of a plastics test specimen should be expressed only as the modulus of the material as manifest in that particular test piece.

3.2

Apart from the limitation embodied in Section 3.1, the geometrical relationships between the modulus of a plastics material and the stiffness of a structure made from it are determined by the laws of mechanics and are identical to those that emerge in the theories of elasticity, plasticity, etc. The relatively large deformations or strains that may develop in some plastics test specimens and service items in some situations may require the use of higher order terms in the elastic equations, the imposition of special constraints on the test configurations and even redefinitions of strain, see Supplement S3.2.1. Also, local yielding may occur at stress concentrations, e.g. at loading points, thereby requiring adjustments to the calculations or further constraints on the test configurations.

Some of the linear elastic relationships commonly used in the plastics industry's measurement of modulus are given in Supplement S3.2. There are many such formulae, but most of them were developed initially as a means of calculating the stiffness of engineering structures when the elastic properties of the material of construction are known. However, when attempting to derive a modulus value from a practical measurement of stiffness only a few of the configurations are suitable for the accurate determination of a modulus value.

Initially, elastic relationships commonly used in the metals engineering industry were adopted without modification for the derivation of the 'moduli' of plastics but even the earliest experiments showed that those moduli vary with the duration of continuous stressing and the frequency of cyclic stressing. An elementary approach to that feature regards the corresponding viscoelastic functions as merely 'time-dependent' elastic coefficients, but the reality is much more complex than that. In some cases that time-dependence can severely complicate the computations, see Eq. 3.3.1. However, the geometrical aspect is often unaffected, except when the time-dependent component of the deformation becomes large enough to introduce non-linearities into the governing equation or to induce local distortions of the strain field, e.g. stress concentrations at loading points.

3.3

Elasticity theory relates stress to strain via elastic constants, but viscoelasticity theory incorporates stress, strain and time in a fundamental way, through constitutive equations in which the quantities E, G, etc. of elasticity theory are replaced by functions of time or frequency linked by convolution integrals, see Supplement S3.3. When the time dependence is slight, and when no second viscoelastic quantity is involved, the equations are simply modified by the replacement of E, for example, by $E(t)$. Otherwise, the elastic equations have to be replaced by their viscoelastic analogues, see Eq. 3.3.1 for example,

$$E(t) = 2G(t)[1 + v_0] + 2\int_0^t G(u)v(t-u)du \qquad [3.3.1]$$

where $E(t)$ = time-dependent relaxation modulus in tension
   $G(t)$ = time-dependent relaxation modulus in shear, and
   $v(t)$ = time-dependent relaxation lateral contraction ratio.

Such mathematical elaboration is necessary in measurement technology and in engineering design for plastics materials only when the strain rates are high but the usual simple process of substituting $E(t)$ for E in the elastic equations has tended to engender a mistakenly casual attitude as to how the $E(t)$ should be measured.

3.4

Irrespective of the deformation mode, the moduli may be obtained from three classes of excitation, i.e.

- constant stress rate (or strain rate),
- sinusoidally varying stress (or strain),
- abruptly applied stress (or strain) held constant thereafter,

which are, respectively, the ramp, sinusoid and step excitation procedures referred to in Chapter 2. Those particular excitations are relatively easily handled by viscoelasticity theory and by practical mechanisms.

There is consequently a profusion of definitions. For example, for the tensile mode of deformation there are the following common designations:

- relaxation modulus
- creep modulus
- complex modulus (with real part and imaginary part)
- tangent modulus (from a traditional ramp excitation tensile stress–strain curve)
- secant modulus (from a traditional ramp excitation tensile stress–strain curve)

with additional, subsidiary classifications such as:

- relaxed, unrelaxed, short-term, long-term, etc.

and information about the test conditions, such as, for instance, $E_r(0.002, 100, 23)$ signifying a relaxation modulus in tension 100 seconds after the application of a strain of 0.002 at a temperature of 23 °C.

3.5

If the material is isotropic and linear viscoelastic, the various moduli are related to one another through linear integral equations such as Eq. 3.3.1, which are analogues of the equations of elasticity theory. In principle for these conditions, data from different classes of experiment, i.e. the responses to different excitation functions can be combined into a single relationship between modulus and log elapsed time or log frequency, at any one temperature. However, plastics are non-linear viscoelastic, i.e. the stress is not linearly proportional to the strain, see Supplement S3.4, and, furthermore, they are seldom isotropic after they have been processed into end-products. Therefore, the relationships between the moduli are correspondingly more complex than those of the isotropic linear viscoelasticity theory and largely intractable by analytical methods.

3.6

If the test specimen/item is severely anisotropic, additional nomenclature is required, e.g. 'longitudinal modulus' and 'transverse modulus', or adoption of the stiffness and/or compliance coefficients, see Chapter 7. The latter is the preferred nomenclature because its formal correctness reduces the possibility of ambiguities in the interpretation of results, but it is fully descriptive only when the state of order does not vary over the gauge region, which is rare in injection mouldings. Those coefficients are material properties with the same status as the moduli of the isotropic case and are not to be confused with the stiffness of an article (Section 3.1).

3.7

The modulus value quoted in a typical elementary data sheet is usually obtained by the use of a ramp excitation (i.e. the first of the three classes of test cited in Section 3.4) and, for various reasons, it will usually differ from values generated by other methods. Typically, the effective value in a service situation is significantly lower than the commonly tabulated value; the combined effects of high strain, long elapsed time under load and unfavourable molecular alignments might cause the effective modulus to be as low as one tenth of the commonly published value. That effective downrating of modulus causes serious difficulties for data generators and data utilizers alike. An effective database needs to reflect the multivalued nature of modulus (and other properties) and to incorporate an expert system to advise on the matching of a datum from the array of possible values to the requirements of particular situations. There is increasing scope for error (the use of inappropriate data) as data handling processes such as design calculations are automated.

3.8

In summary, the nature of plastics is such that any modulus

- decreases with increasing temperature;
- decreases with increasing strain;
- decreases with increasing time under stress;
- varies with the thermal history of the sample;
- probably varies from point to point in a moulded item;
- probably varies with the axis of applied stress.

The form and the degree of the time dependence and the other effects is governed by the molecular architecture of the polymer and the molecular state of the particular sample under investigation. It follows that the subjective impression that a particular plastic has a 'high' modulus or a 'low' modulus depends on the time-scale of the experiment in relation to the shape and location of the characterizing modulus vs. log time and modulus vs. temperature relationships (see also Section 3.10), on how effective the crystallinity is in diminishing the mechanical effects of the glass–rubber transition and on how other aspects of molecular order modify the shape of $M = f(\log \text{time}, T)$.

3.9

The commonly used designation 'short-term' denotes that a property value has been measured shortly after the imposition of an excitation or even during its imposition (for example, during the course of a conventional stress–strain test). It is an imprecise definition that carries various implications depending on the plastic and the circumstances. For example, the short-term modulus of a plastic in its glassy state is almost independent of the elapsed time or the strain rate and therefore it has a descriptive significance whereas the short-term modulus of a plastic in or near its glass–rubber transition region is sensitive to the elapsed time and is correspondingly not definitive, though such data nevertheless quantify the subjective impressions gained by casual handling.

The tensile modulus lies in the range 100 kPa to 1 MPa for a plastic in its rubbery state, in the range 2.5 GPa to 5 GPa for a plastic in its glassy state, and in the range 1 MPa to 1 GPa for amorphous plastics in their glass–rubber transition region and for crystalline plastics above their T_g. High values up to about 10 GPa are observed at very low temperatures. The comparable modulus of a steel is 200 GPa. Some comparative data are given in Table 3.9/1, from which one can infer the considerations that arise when plastics are substituted for metals in load-bearing situations.

3.10

The combined effects of elapsed time and temperature on the modulus are best represented by a three-dimensional model, but an alternative two-dimensional contour map shows the locus and abruptness of the modulus transitions (secondary, glass–rubber and, where relevant, crystal melting) on a temperature–log time map.

The transition map can be so constructed as to span a very large range of log time by use of the viscoelastic equations that relate the separate responses to step excitations and sinusoidal excitations. The basic relationship is:

$$C*(\omega) = \int_0^\infty C_c(t)e^{-j\omega t}\,dt \qquad\qquad [3.10.1]$$

where $C*(\omega)$ = complex compliance
and $C(t)$ = creep compliance

See Supplement S3.3 and Chapter 5 for elaborations of those definitions.

Table 3.9/1 Approximate values of tensile modulus and density (20 °C)

Material	Young's modulus (GPa)	Relative density
Carbon steel	200	7.8
Rolled copper	120–130	8.9
Cast iron	85–100	~7.7
Cast aluminium	55–75	2.7
Polyester resin + 87% w/w uniaxially oriented glass fibres	52	2.2
Polyester resin + 60% w/w glass cloth	21	1.7
Polyester resin + 30% w/w glass fibre material	10.5	1.3
Nylon 66 + 33%w/w glass fibres (dry)	10	1.4
Oak (dry)	~10	~0.7
Polymethyl methacrylate	3	1.2
Propylene homopolymer (sp. density 0.907)	1.5	0.9

The data in this table must be treated with reserve and as nothing more than rough indicators. The properties of metal alloys depend on the composition and the treatment, those of plastics depend on ambient temperature, time under load, strain, processing history, etc. (the values quoted here refer to short times, small strains and favourable flow geometries). The modulus of oak is similarly dependent on several variables.

The literature of the 1950s abounds with transition maps and discussions of the mathematical niceties. Such matters were highly pertinent at that time, guiding and underpinning the exploratory chemistry of the period that produced many of the polymers in common use today, but the downstream demands are now different, often requiring a higher precision and accuracy in property measurements than is possible via linear viscoelastic mapping. The outcome of earlier mapping activities, largely relegated to the archives, now plays the useful but secondary role of indicating which regions of time and temperature are likely to be the most rewarding during contemporary test programmes and what inferences might be drawn about the modulus in a neighbouring region.

3.11

The precision and accuracy with which the modulus of a single specimen can be measured depends on the mechanical details of the test configuration. The principal sources of inaccuracy are:

1   the stress state in the critical region deviating from that assumed in the theory;
2   the strain not being uniform over the critical region;
3   the assumed stress–strain relationship being inadequate, e.g. due allowance not being made for anisotropy and non-linearity;
4   inadequate control of loading path and ambient temperature;
5   deviations from the norm in the state of order in the specimen;
6   variations in the state of order from point to point in the specimen.

The principal sources of imprecision are random variations in the state of molecular order in the set of test specimens, in the critical elements of the test configuration and in the test procedure.

Leaving genuine interspecimen variability aside, careful replicate tests for precision in measurements of tensile modulus typically have a coefficient of variation (c.v.) of 0.02 or less. For modulus in flexure, the c.v. for a set of nominally identical tests should be no more than 0.01. Those low levels of variation are possible only through impeccable experimentation and are not attainable in measurements of complex compliance and therefore some degree of mismatch is likely when $C^*(\omega)$ data and $C(t)$ data are combined via Eq. 3.10.1.

3.12

A general guiding principle for the measurement of modulus is that the applied stress or strain should be simple, i.e. uniaxial tension, beam flexure, etc., and that has governed the structure of the sections and the supplements of this chapter. However, many practical load-bearing situations entail combined stresses, e.g. tension combined with torsion, and sometimes such situations are simulated in laboratory tests. In those cases, the terms equivalent stress and equivalent strain be usefully deployed, see references 1 and 2 and others. In general, however, where precision and accuracy in modulus measurement are paramount it seems that good practice requires the avoidance of deliberate or fortuitous combined stresses.

References

1 P. D. Ewing, S. Turner and J. G. Williams. 'Combined tension–torsion studies on polymers: apparatus and preliminary results for polythene.' *Journal of Strain Analysis* 7, 1, 9, 1972.
2 P. D. Ewing, S. Turner and J. G. Williams. 'Combined tension–torsion creep of polyethylene with abrupt changes of stress.' *Journal of Strain Analysis* 8, 2, 83, 1973.

S3.2   Modulus – linear elastic solutions

S3.2.1 Definitions of strain

Strains are defined as follows:
Engineering strain

$$\varepsilon = \frac{\delta l}{l_o}$$ [S3.2.1.1]

Extension ratio

$$\lambda = \frac{l}{l_o} = \varepsilon + 1$$ [S3.2.1.2]

Logarithmic strain (Henky measure)

$$\varepsilon_H = \ln \frac{l}{l_o} \text{ or } \ln \lambda$$ [S3.2.1.3]

In combined stress and strain situations there is also equivalent strain:

$$\varepsilon_{equiv} = \frac{\sqrt{2}}{3}\left[(\varepsilon_{xx} - \varepsilon_{yy})^2 + (\varepsilon_{yy} - \varepsilon_{zz})^2 + (\varepsilon_{zz} - \varepsilon_{xx})^2 + 6\left(\varepsilon_{xy}^2 + \varepsilon_{yz}^2 + \varepsilon_{zx}^2\right)\right]$$

[S3.2.1.4]

where

> δl is increase in length
> l_0 is an original gauge length
> l is an increased length
> ε_{ij} is strain in the plane i, j with i, j representing the rectangular coordinates x, y, z

Strain may be defined in whichever way is convenient for the solution of a particular problem. ε is useful when the strains are small, λ is useful when the strains are large, ε_H is versatile, e.g. if a rod is stretched to twice its original length or reduced in axial compression to half its original length the value of ε_H is 0.694 or −0.694, respectively. Equivalent strain is a useful concept for combined strain situations.

S3.2.2 Flexure of beams

If an isotropic, elastic, homogeneous, long, thin and narrow beam of uniform cross-section is subjected to a transverse force and if the resulting deflection is less than about half its thickness the elastic response of the beam is given by:

$$\delta = \frac{PL^3}{AEI} \qquad\qquad\text{[S3.2.2.1]}$$

where δ = deflection of beam
   P = applied lateral force
   L = effective length of beam, i.e. the span
   I = area moment of inertia
   E = modulus of elasticity
   A = a shape factor (see details in Table S3.2.2/1).

Many formulae such as this are used by the engineering community to calculate the likely deformation of structures subjected to external forces when the value of the relevant modulus is known. Some of the equations, including S3.2.2.1, can be used in the reverse sense to calculate the value of the modulus from a measurement of the deformation. Thus, for instance, provided the force is applied as an abrupt step and provided the immediate response is disregarded the modulus (as a function of time) can be derived from the following variant of Eq. S3.2.2.1:

$$E_c(t) = \frac{PL^3}{AI\delta(t)} \qquad\qquad\text{[S3.2.2.2]}$$

where $E_c(t)$ = tensile creep modulus.

Table S3.2.2/1 Shape factors for transversely
loaded beams

Manner of support and loading	A
Cantilever, end load	3
Cantilever, uniformly distributed load	8
Freely supported, centre load	48
Freely supported, uniformly distributed load	$\dfrac{384}{5}$
Fixed ends, centre load	192
Fixed ends, uniformly distributed load	384
Fixed at one end, freely supported at the other, centre load	107
Fixed at one end, freely supported at the other, uniformly distributed load	187

Some of these configurations are not suitable for
the measurement of modulus.

Table S3.2.2/2 Area moments of inertia

Cross section	I
Rectangular, width b, depth h	$\dfrac{bh^3}{12}$
Solid circular, diameter h	$\dfrac{\pi h^4}{64}$
Hollow circular, outer diameter h_1, inner diameter h_2	$\dfrac{\pi(h_1^4 - h_2^4)}{64}$

The tensile creep modulus is the reciprocal of the tensile creep compliance and is not strictly equivalent to the relaxation modulus, though the
approximation is close when the time dependence is slight and the strain is
so small that the material is virtually linear viscoelastic. See Chapters 5 and
6 for details of the time forms of the common excitations.

The shape factor in Eq. S3.2.2.1 depends on the test configuration. Some
values are given in Table S3.2.2/1.

Area moments of inertia, I, for simple cross-sections are given in Table
S3.2.2/2.

High accuracy is attainable from the use of Eq. S3.2.2.1 only if restrictions are placed on the relative dimensions of the specimens to eliminate
certain second-order effects that have been recognized for many years, see

correspondence in reference 1. Additionally, the equation assumes that the Young's modulus in tension is equal to that in compression, but there has been some uncertainty as to whether that is so in the case of plastics except at infinitesimal strains. That issue has not been easy to resolve, but the use of a macrocomposite specimen in which a thin lamina of plastic was bonded to a thicker lamina of steel such that during flexure the plastic component was entirely under either tension or compression suggested that the two moduli are virtually identical for many plastics.[2] That formula is based on the assumption that the modulus does not vary from point to point in the beam; when the modulus varies through the thickness, as it often does in a plastics specimen because of variations in the state of molecular order, that variation cannot be quantified by the straightforward flexed beam method, nor can the notional value of E that is then derived via equation S3.2.2.1 be regarded as a mean value because of the way in which the flexural stiffness is dominated by the skin layers. In a tension test the properties of the individual lamellae contribute in parallel and without bias to the overall property and the force–deflection curve relates to an averaged value of the modulus. In contrast, in a flexure test the contribution of each lamella depends on its disposition with respect to the neutral axis and hence the datum generated in the test may be a notional modulus related to the stiffness of the beam rather than an accurate measurement of the modulus of the material.

Apart from those assumptions, Eq. S3.2.2.1 accounts only for deformation due to bending but there is also a component due to shear. The shear stress on a vertical section varies from 0 at the surfaces to a maximum of $3P/4bh$ at the neutral plane and if uniformly distributed it would be $P/2bh$. That shear stress generates a shear strain in the beam which contributes to the overall deflection, i.e.

$$\delta = \delta_b + \delta_s \qquad\qquad [\text{S3.2.2.3}]$$

where δ_b = the deflection of Eq. S3.2.2.1

$$\text{and} \quad \delta_s = \frac{PL}{4bhG} \qquad\qquad [\text{S3.2.3.4}]$$

The modified equation for an isotropic specimen subjected to 3-point flexure is:

$$\delta = \frac{PL^3}{4bh^3 E}\left(1 + \frac{Eh^2}{GL^2}\right) \qquad\qquad [\text{S3.2.2.5}]$$

where δ = deflection at the mid-point
P = force at the mid-point
L = span

b = width
h = thickness
E = Young's Modulus (appropriate to the particular anisotropy)
G = shear modulus.

It follows from Eq. 3.2.2.5 that the shear component may be important if the beam is thick and short or if the ratio E/G is large, as may arise in foamed-core sandwich beams and in certain laminates. A set of tests on beams of identical cross-section but different lengths (spans) enables E and G to be derived (δ/PL vs. L^2 provides E from the slope and G from the intercept; δ/PL^3 vs. $1/L^2$ provides G from the slope and E from the intercept). See also Supplement S7.1.3.

The product EI is known as the flexural rigidity, designated D, and the product bhG is known as the transverse shear stiffness, designated Q. Equation S3.2.2.5 can be rewritten as:

$$\delta = \frac{PL^3}{48D}\left(1+\frac{12D}{L^2Q}\right)$$

[S3.2.2.6]

for simplification in structural calculations. For example, shear is considered to be negligible when:

$$D/L^2Q < 0.01$$

and shear deflections are comparable to bending deflections when:

$$0.01 < D/L^2Q < 0.1.$$

See reference 3 for further details.

If the beam is wide, E in Eq. S3.2.2.1 has to be replaced by $E/(1 - v^2)$, which is often designated C and called the stiffness. However, if such a plate is subjected to three-line flexure, but to a deflection no greater than, say, one-half of the thickness and that test is repeated as the width is reduced progressively the derived stiffness coefficient decreases from what is an approximation to $E/(1 - v^2)$ to E, see Fig. 8 in reference 4, for example, and Fig. S3.2.2/1, which relates to a laminated plate of poly(ether sulphone) and twill weave glass fibre mat.[5] Both sets of results imply that modulus in flexure should be carried out either on very narrow beams, for the derivation of E, or on wide plates, for the derivation of $E/(1 - v^2)$. The limiting values of the width/thickness ratio that define 'narrow' and 'wide' depend on the degree of anisotropy of the specimen, but it is clear that the use of intermediate ratios gives unspecific data presumably because anticlastic curvature is important in that range.

The modulus value derived from three-line flexure of a plate equates to the mean value of the modulus of a set of narrow strips cut from the plate and since the modulus may vary with the state of molecular order from

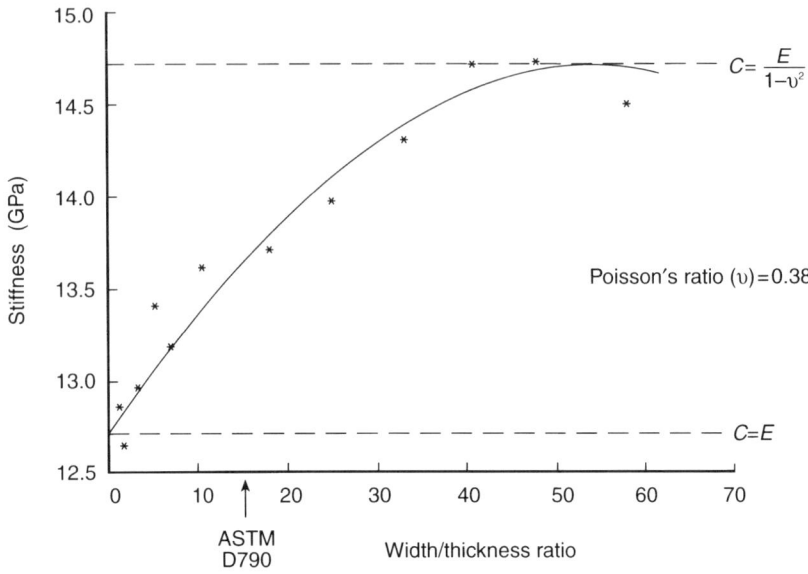

$S3.2.2/1$. The stiffness of wide flexed plates. The influence of the width/thickness ratio on the stiffness of a PES/glass fibre mat laminate.

point to point in the plate it may be a more representative value than that of a particular beam. Three-line flexure of a square plate successively in orthogonal directions gives information about the level of its modulus anisotropy, see Supplement S3.2.3 and Chapter 7.

References

1 Letters (E. A. Richardson, S. Timoshenko.) *Mechanical Engineering* 45, 259, 1923.
2 H. Kondo and T. Kuroda. 'On the moduli of elasticity in bending on plastics'. Proceedings of the Ninth Japanese Congress on Testing of Materials.
3 A. F. Johnson and G. D. Sims. 'Stiffness properties of sandwich foam materials.' NPL Report DMA (D) 153, December 1978.
4 R. C. Stephenson, S. Turner and M. Whale. 'The assessment of flexural anisotropy and stiffness in thermoplastic-based sheet materials.' *Plastics and Rubber: Processing and Applications* 5, 7, 1980.
5 D. C. Leach. Private communication.

S3.2.3 Flexure of plates

There is a wide variety of combinations of plate geometry, edge constraints and loading for plates but the solutions are of the common form:

$$\delta = \frac{3Pa^2 f(v)}{4\pi Eh^3}$$
[S3.2.3.1]

where P = applied force
a = a lateral dimension
$f(v)$ = a function of Poisson's ratio
E = Young's modulus
h = thickness

That equation is valid if:

1 the shear stresses are small;
2 membrane stresses are absent or insignificant;
3 the plate is flat and of uniform thickness;
4 there is no yielding of the material and no surface tractions at the support;
5 the material is homogeneous, isotropic and linear elastic;
6 the modulus in tension is equal to that in uniaxial compression.

To attain those conditions it is necessary that:

1 the maximum deflection should be less than half the thickness;
2 the radius of the support should be greater than four times the thickness;
3 the line of action of all the forces should be normal to the plane of the plate.

The commonest service situations for circular plates are:

Central load, freely supported edge, for which $f(v) = (1 - v)(3 + v)$;
Central load, clamped edge, for which $f(v) = (1 - v^2)$;
Uniformly distributed load, freely supported edge, for which $f(v) = (1 - v)(5 + v)/4$;
Uniformly distributed load, clamped edge, for which $f(v) = 2.5(1 - v^2)$.

The formulae are mainly intended for use in design calculations, to determine section thicknesses etc. when the modulus is known, not for the reverse process of measurement of modulus. However, there are certain circumstances in which tests on entire plates can be an economic route to useful data. Clamped plates are generally unsuitable for the purpose because the edge condition can never be established precisely. The third configuration has been used on one occasion,[1] but is impractical for general use and only the first configuration is a reasonably practical option. A centrally-loaded, edge-supported disc is the simplest test configuration and is widely used for the measurement of impact resistance (see Chapter 11). In principle, it should provide a modulus value that is more effective as a characterizing datum than one derived, say, from a simple beam cut from the plate if the plate is anisotropic. However, the results from various trials have never been encouraging, probably because the test pieces were either not perfectly flat initially or distorted asymmetrically under load because of anisotropy; in either case the reaction at the support varies around the periphery, changing the boundary conditions. Pouzada and Stevens[2] avoided the out-of-flatness problem by using 3-point support, which seems to be a practical and superior alternative method; their shape factor, $f(v)$, is rather more complicated than those given above, but not unmanageable and not seriously sensitive to the value of v.

Those configurations have the theoretically attractive advantage of axial symmetry but it transpired that the slightly unorthodox configuration of three-line flexure of a disc is very informative, particularly when the sample/test piece is anisotropic, which is often the case. The matter is discussed in Chapter 7.

References

1 D. McCammond and S. Turner. 'Poisson's ratio and the deflection of a viscoelastic Plate.' *Polymer Engineering and Science* 13, 3, 187, 1973.
2 A. S. D. Pouzada and M. J. Stevens. 'Methods of generating flexure design data for injection moulded plates.' *Plastics and Rubber: Processing and Applications* 4, 2, 181, 1984.

S3.2.4 Torsion and compression

Linear elastic analysis for other modes of deformation can also be used in order to determine modulus, for example, axial modulus by compression and shear modulus by torsion. Uniaxial compression of a cylinder or plate type specimen are commonly used to determine either modulus or some specific stress, e.g. at yield. These are discussed in Chapters 4 and 6 in Supplements S4.7.2 and S6.5.

Torsion of a rectangular prismatic beam, at small deformations, is popular for the determination of shear modulus for isotropic and transversely isotropic samples. This is discussed in detail in Chapter 7, Supplement S7.1.4. Torsion of a square or rectangular plate can be used to determine an in-plane shear modulus. The simplest approach is with a square plate,[1] which is subjected to pure bending by the action of four point loads. Two equal upward forces are applied at the ends of one diagonal and two equal but opposite forces at the ends of the other diagonal. The effective plate shear modulus (in-plane shear modulus) can be calculated from the slope of the force–crosshead displacement curve (plate twist stiffness):

$$G = \frac{3}{8}\frac{L^3}{h^3}\left(\frac{P}{\delta}\right)$$

[S3.2.4.1]

where

L is the length of the diagonal
P/δ is the plate twist stiffness
h is the thickness of the plate.

The plate is bent into an anticlastic surface. The equation only applies for small deflections and also for small loads, since there is a requirement to ensure that the response is pure shear (satisfied by a thin plate with a length to thickness ratio of 40). This is not of concern for its adoption as a test method but might be a limitation in a design context where restrictions on practical deformations may be inappropriate.

Equation S3.2.4.1 applies to isotropic and transversely isotropic plates. For example shear modulus of a polymethylmethacrylate (PMMA) sample determined from beam torsion agrees with in-plane shear modulus of the same PMMA sample determined by plate twist. However, if these approaches were applied to, say, injection moulded discontinuous fibre reinforced composite plates then agreement between shear modulus and in-plane shear modulus might not occur. For anisotropic samples in the form of plates, there exists an asymmetry within the plane which introduces a flexural component to add to the pure shear. The plate stiffness is then increased.

Other approaches[2] with the plate twist test have broadened its application to rectangular plates (side dimensions a_1 and a_2). In addition, as well as loading at the corners it is possible to load at four symmetric positions within the plate boundaries (sometimes considered to be more convenient). A slightly different expression is then used to determine the in-plane shear modulus:[2]

$$G = \frac{3}{4}\left(\frac{P}{\delta}\right)\frac{Ea_1a_2f(a)}{h^3}$$

[S3.2.4.2]

where E is an axial modulus and $f(a)$ is the factor to allow for the position of the loading points.

References

1 R. C. Stephenson, S. Turner and M. Whale. 'The load-bearing capability of short-fibre thermoplastics composites – a new practical system of evaluation.' *Polymer Engineering and Science* 19, 3, 173, 1979.
2 W. Nimmo and G. D. Sims. 'Composite materials data for structural beam design'. NPL report DMM(D)88, Nov 1991.

S3.3   Linear viscoelasticity

The general phenomenological theory of linear viscoelasticity relates any time-varying excitation function (of either stress or strain) to the corresponding response function (of either strain or stress). The central equation states that stress, strain and time are related via the differential equation:

$$a_n \frac{\partial^n \sigma}{\partial t^n} + \ldots + a_1 \frac{\partial \sigma}{\partial t} + a_0 \sigma = b_m \frac{\partial^m \varepsilon}{\partial t^m} + \ldots + b_1 \frac{\partial \varepsilon}{\partial t} + b_0 \varepsilon \qquad [S3.3.1]$$

from which the entire mathematical structure for the comprehensive representation of linear viscoelasticity develops. It is the basis of the mechanical testing of plastics even when the actual behaviour deviates from linear.

The coefficients $a_0 \ldots a_n$ and $b_0 \ldots b_m$ have the status of materials constants. Equation S3.3.1 reduces to an elastic expression when all of the coefficients other than a_0 and b_0 are zero:

$$\frac{\sigma}{\varepsilon} = \frac{b_0}{a_0} = E \qquad [S3.3.2]$$

However, when some of the other coefficients are not zero the extraction of a modulus is not so simple. The difficulty is resolved by Laplace transformation of Eq. S3.2, which enables a 'transform compliance' to be defined thus:

$$\overline{C} = \frac{\overline{\varepsilon}}{\overline{\sigma}} = \frac{(a_n p^n + a_{n-1} p^{n-1} \ldots \ldots + a_0)}{(b_m p^m + b_{m-1} p^{m-1} \ldots + b_0)} \qquad [S3.3.3]$$

and similarly its reciprocal, the transform modulus. There is a set of such transform compliances and transform moduli which correspond completely to the elastic compliances and moduli of an elastic material, with the relationships between the transform moduli corresponding exactly to those between the elastic moduli. The relationships between the untransformed viscoelastic moduli and compliances, on the other hand, are not identical to their elastic counterparts; for example, the relaxation modulus (stress response divided by constant applied strain) is not the reciprocal of the creep compliance. That particular result follows from Eq. 3.3.3 and the definitions of relaxation modulus and creep compliance. The latter is:

$$C_C(t) = \frac{\varepsilon(t)}{\sigma_0} \qquad [S3.3.4]$$

where $\varepsilon(t)$ = time-dependent strain response
σ_0 = instantaneously applied constant stress.

and from Eq. S3.3.3,

$$\bar{\varepsilon} = \frac{(a_n p^n + a_{n-1} p^{n-1} \ldots\ldots + a_o)}{(b_m p^m + b_{m-1} p^{m-1} \ldots + b_o)} \bar{\sigma}$$

so that $\bar{C}_c(t) = \dfrac{\bar{\varepsilon}(p)}{\sigma_o} = \dfrac{\bar{C}}{p}$

and $\overline{M}_r(t) = \dfrac{\bar{\sigma}(p)}{\varepsilon_o} = \dfrac{\overline{M}}{p}$

where $\overline{M}_r(t)$ is the Laplace transform of the relaxation modulus and $\bar{C}_c(t)$ correspondingly for the creep compliance,
whence

$$p\bar{C}_C(p) = \frac{1}{p\overline{M}_r(p)} \tag{S3.3.5}$$

and the relationship between relaxation modulus and creep compliance is given by the convolution integral:

$$\int_0^t M_r(t)C_c(t-u)du = t \tag{S3.3.6}$$

Under some circumstances, relationships such as Eq. 3.3.6 collapse into the elastic counterparts, and at other times their influence is circumvented by the simple expedient of segregating the test methods and the results arising therefrom. It is helpful also if the excitation is so prescribed that most or all of its time derivatives quickly attain a steady state, so that the relationship between a modulus (or compliance) as measured by the procedure bears a simple relationship to the transform modulus (or compliance), which is the fundamental quantity. The complex modulus tests are ideal in that respect (see Chapter 5) and the step excitation methods are relatively simple, as shown by Eq. 3.3.5; others are best avoided, though the ramp excitation method, see Chapter 4, which entails a higher level of complication enjoys a long-standing popularity, probably because the results can be used largely within the constraints of a segregated class.

If either $M_r(t)$ or $C_c(t)$ can be represented analytically it may be possible for it to be transformed into the other function. Thus, for instance, if:

$$M_r(t) = a\left(\frac{t}{b}\right)^{-m} \tag{S3.3.7}$$

it can be shown[1] that

$$C_c(t) = \frac{\sin mp}{mp} \frac{1}{M_r(t)} \tag{S3.3.8}$$

The adjustment term $\dfrac{\sin mp}{mp}$ approximates to unity for small m thus:

m	$\dfrac{\sin mp}{mp}$
0.05	0.996
0.10	0.984
0.20	0.935

The functions $C_c(t)$ and $M_r(t)$ are particularly important in the theory of linear viscoelasticity because they characterize the response to step excitations and hence, through a superposition rule for linear systems, the response to any arbitrary excitation. Boltzmann's Superposition Principle states that, for a linear viscoelastic material, the strain response to an arbitrary stress excitation function, $\sigma(t)$, is given by:

$$\varepsilon(t) = \int_0^t C(t-u)\frac{d\sigma(u)}{du}\,du \qquad\qquad [\text{S3.3.9}]$$

where $C(t)$ = the creep compliance function.

There are several alternate forms of Eq. S3.3.9, which may be derived by integration by parts and change of variable.

Via the superposition principle, the creep compliance is related to the complex compliance by:

$$C*(\omega) = \int_0^t C_c(t)e^{-j\omega t}\,dt \qquad\qquad [\text{S3.3.10}]$$

The various superposition integrals may be derived analytically provided that the response to an additional excitation is independent of, and linearly addable to, any prevailing response to previously imposed excitations (i.e. provided that the system is linear). They can also be derived graphically, but both methods are prone to serious cumulative errors that arise from even small deviations from linearity and virtually unavoidable experimental imprecisions and interspecimen variabilities; see Chapter 6 for instance.

Reference

1 A. A. McLeod. 'Design of plastic structures for complex static stress systems.' *Industrial Engineering and Chemistry* 47, 1292, 1955.

S3.4   Non-linear viscoelasticity

There have been numerous attempts to develop mathematical models of non-linear viscoelasticity, ranging from the use of non-linear components in the simplest spring-dashpot models, such as the Maxwell and Voigt elements, to the multiple integral representation. A typical intermediate formulation is the Leaderman expression, proposed long ago, which replaces the linear superposition integral:

$$\sigma(t) = \int_0^t M_r(t - u)\frac{d\varepsilon}{du}\,du \qquad\qquad [S3.4.1]$$

with
$$\sigma(t) = \int_0^t M_r(t - u)\frac{d}{du}(f(\varepsilon))du \qquad\qquad [S3.4.2]$$

That equation and the many subsequent variants of it had some strictly limited success in representing the response to time-varying excitations if each imposed change of stress or strain was an increase but they failed to represent the response to decreasing excitation, see also Chapter 6. The explanation for the general inadequacy of the superposition equations now seems to be that each change of stress or strain alters the equilibrium state of molecular order over a period of time, the specimen tending to drift towards the new equilibrium as it simultaneously responds to the changed excitation i.e. the inherent modulus function, $M_r(t)$, or compliance function, $C(t)$, changes at any time u_1, etc. at which the stress is changed. Thus, it is questionable whether the response to a stress change will ever be quantifiable via non-linear viscoelasticity theory, mainly because crucial facts are not derivable from the relaxation or creep experiments. Apart from that fundamental constraint, the structure of the theory is such that the critical observable quantities from which the modifying functions might be derivable cannot be measured sufficiently accurately, because of the limitations of the apparatus and inevitable interspecimen variability. On the other hand, within the limits of attainable accuracy and precision, simple patterns of behaviour do emerge, as is shown in Chapter 6, at least for some plastics, and the response to various stressing histories can be predicted approximately by judicious extrapolation of, or interpolation between, experimental data. However, satisfactory outcomes in this field rely at present much more on the practical skills and wisdom of the experimenter than on his/her prowess with the theories and whilst the results can suffice for many practical purposes they do not suffice, at present, to support emerging theories.

Modulus from constant deformation rate tests

4.1

Tests under conditions of constant rate of straining or stressing have been widely used for the mechanical evaluation of many classes of material long before they were adopted by the plastics community. For a viscoelastic material they constitute the class of ramp excitation tests, see Chapter 3.

When applied to plastics, ramp excitation is primarily used to measure strength, for which it is eminently satisfactory. However, during the initial stages of the ramp it imposes an approximately constant straining rate on the specimen and with appropriate instrumentation short-term secant moduli and tangent moduli can be derived from the force–deformation curve, see Fig. 4.1/1. For plastics materials those moduli are multivalued because the force–deformation and stress–strain relationships vary with the deformation rate and the ambient conditions and additionally are not linear.

The commonest use of ramp excitation is for the conventional tensile test, which is one of the oldest mechanical test procedures, so much so that ramp excitation is almost synonymous with tensile test, but flexure, uniaxial compression and even shear are readily attainable, as will emerge later.

4.2

Modulus was defined originally within the framework of linear elasticity. Curvature of the force–deformation or stress–strain relationship, such as that shown in Fig. 4.1/1, is indicative of stress–strain non-linearity if the material is elastic. However, it is not unambiguously indicative of it if the material is viscoelastic, because with a constant deformation rate test, both strain and time vary along the abscissa and both variables exert an influence on the shape of the response. For viscoelastic materials the only unambiguous indicator of system non-linearity is curvature on an isochronous stress–strain curve (Chapter 6).

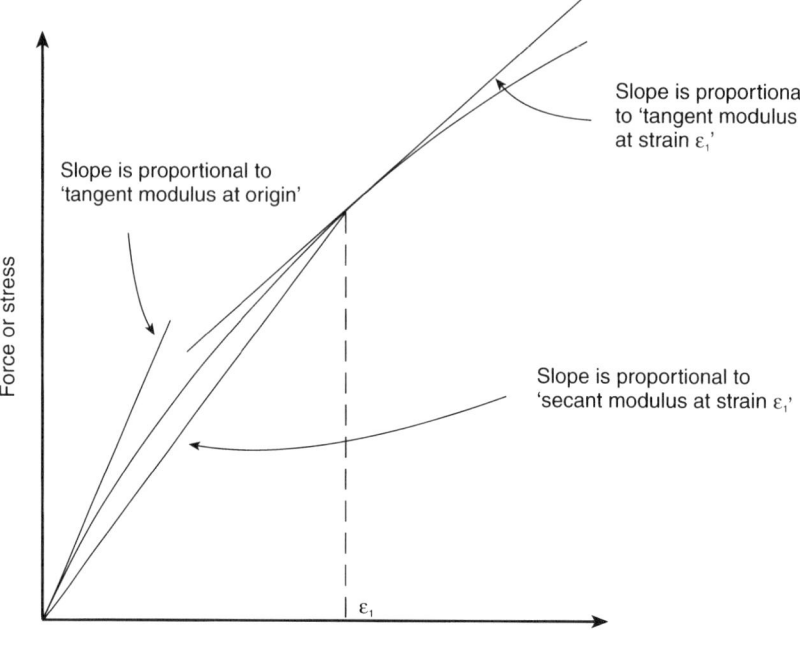

Force or stress

Slope is proportional to 'tangent modulus at origin'

Slope is proportional to 'tangent modulus at strain ε_1'

Slope is proportional to 'secant modulus at strain ε_1'

ε_1

Deflection or strain

4.1/1. Force-deformation relationships below the yield point. The tensile modulus of most plastics materials decreases as the strain increases. The curvature is due to viscoelasticity and possibly structural change during the test.

4.3

Tensile measurements of strength are simple to conduct and the results so generated are reliable, though not necessarily easy to interpret and not even meaningful if failure occurs far beyond the yield point. Tensile tests intended to measure modulus are more troublesome in that they should entail the use of an extensometer because actuator movement does not translate accurately into specimen strain, though for some purposes it may suffice if the latter is used as an approximation to displacement, as was the usual practice until relatively recently. Experimental results are prone to errors arising variously from relative movements and/or stress concentrations at the test piece-grip and test piece–extensometer interfaces, deviations from axiality due to machine-test piece misalignment and test piece asymmetry, imperfect extensometry and initially varying extension rate due to mechanical inertias, see Supplement S4.3.

4.4

Modern machine designs and data processing routines have combined to limit the errors attributable to the causes listed in Section 4.3 but even so accumulated machine errors combined with interspecimen variability result in coefficients of variation in the range 0.05–0.10 for measurements of modulus by any ramp excitation method.

Evidence from various sources suggests that the lowest attainable interspecimen variability for modulus measurements by any method is manifest as a coefficient of variation of 0.01, but the magnitude is generally much greater than that, see also Section 4.8. The interspecimen variability arises from adventitious differences in the state of molecular order in a set of test pieces, but the degree of variation depends on the polymer architecture, the source of the test pieces, the production method, conditions of storage, pretreatment etc.

4.5

A variant on the traditional tensile test requires the supplementary use of lateral extensometers and calculation of the volume strain ($\Delta V/V$)

$$\frac{\Delta V}{V} = [(1 + \varepsilon_l)(1 + \varepsilon_b)(1 + \varepsilon_h) - 1] \qquad [4.5.1]$$

where ε_l, ε_b, ε_h are axial, width and thickness strains, respectively. If transverse isotropy is assumed, the thickness and width strains are equal and the expression for the volume strain becomes:

$$\frac{\Delta V}{V} = \left[(1 + \varepsilon_l)(1 + \varepsilon_b)^2 - 1\right] \qquad [4.5.2]$$

where ε_l and ε_b are axial and transverse strains, respectively. However, the assumption that the through-thickness and the across-width contraction ratios are equal is less a reflection of the true situation than an experimental convenience; two lateral extensometers are cumbersome to use and the strain measurement along the through-thickness direction is usually less accurate than that in the across-width direction because test pieces are generally rectangular in cross-section and thin relative to their width. By virtue of the flow geometry of injection-moulded thermoplastics and the lay-up geometry of fibre composites the two lateral contraction ratios are likely to be different and therefore inferences drawn on an assumption of equality should be treated with appropriate circumspection. Additionally, whereas the longitudinal strain datum is an average value for the whole gauge length the lateral strain datum is a local value at one point within the gauge region and may be atypical or erroneous if the test piece is heterogeneous or if the position of the lateral extensometer coincides with a flow irregularity or a misplaced fibre.

Despite the reservations about the experimentation, the derived relationship between volume strain and axial strain can shed light on the deformation mechanisms; for example, an excessive increase in volume strain relative to the axial strain suggests the development of cavitation, crazes or phase separation in two-phase systems. The pioneering studies of those matters were carried out on creep equipment, with which appropriately high precision is attainable, but similar results, albeit at lower precision, can be obtained more expeditiously in constant-rate-of-deformation tests.[1]

4.6

Test procedures have been specified by various standardization bodies, but the recommended deformation rates and dimensions differ. The differences are not sufficiently great to affect results dramatically except when the testing rate and the ambient temperature are such as to place the material in or near a transition region which can affect comparisons between data for different plastics and can distort an order of merit.

Extra precautions have to be taken in the experimentation and the associated Standard Methods impose additional constraints if the test pieces are severely anisotropic and/or heterogeneous, e.g. with continuous fibre reinforced composites; see Supplement S4.6.

4.7

The actuator movement may be utilized to impose flexure, uniaxial compression and other deformation modes via subassemblies. Additionally, the specimen may be a critical basic shape, see Supplements S1.11 and S2.7 rather than a simple bar, beam or strut. For example, it is now common practice for plates to be flexed in various test configurations, but see Chapter 7, Supplement S7.7. Modulus is derived from the force–deflection relationship in accordance with the definitions of Fig. 4.1/1, by means of the appropriate form factor and may then be part of the modulus–ductility datum pair discussed in Chapter 2.

Data from flexed plates are unlikely to be as accurate as data measured on a conventional beam because plates are seldom perfectly flat and therefore under transverse loading they tend to twist as they settle onto the supports, which affects the force–deformation curve and hence the modulus value. On the other hand, the result is an effective mean value of the individual moduli of the set of ideal beams that could have been cut from the plate and that would have reflected the variations and variability of the modulus across the width of the plate. Therefore, for some purposes a modulus datum derived from a flexed plate may be more pertinent than a more accurate modulus datum derived from a flexed beam. Another useful variant in the flexure mode is in-plane flexure of a ring. The ring may be cut from a pipe or a pipe coupling, both of which usually contain structural features that may affect the modulus locally. The arrangement is described and commented upon in Supplement S4.7.1. The mechanical arrangement for the measurement of modulus in uniaxial compression is analogous to that for tensile modulus. The attainable accuracy of measurements tends to be lower than for tension because the test piece has to be relatively short to ensure mechanical stability and friction at the specimen–thrust plate interface distorts the stress field. Ideally, the test piece should be a right, circular cylinder.

Shear measurements entail a more elaborate sub-assembly for both torsion and in-plane shear. For the former the specimen should preferably be a hollow circular cylinder and, failing that, a solid one. Cross-sections other than circular do not remain plane under torsion and the accuracy of the results may be compromised despite the application of a correcting shape factor. The variability and accuracy of torsional data obtained under the best conditions, i.e. hollow cylinders of circular cross-section twisted though only small angles, are similar to those obtained in flexure. The 'rail shear' method is used to measure in-plane shear modulus, only occasionally for plastics but more commonly for fibre-reinforced composites. However, long specimen preparation times are a disincentive and the stress field in the gauge region tends to be distorted by the necessarily high clamping forces. Overall, the plate shear tests are neither accurate nor precise when used for the measurement of modulus; they were designed primarily for strength tests on polymer–fibre composites.

Bulk modulus is rarely measured because the requisite apparatus is complicated and the results are prone to serious errors but it can be measured under ramp excitation, together with uniaxial compression, by an indirect route, see Supplement S4.7.2.

4.8

A measurement of modulus by flexure of a beam generally gives a more precise value than a measurement of tensile modulus, provided that the standard constraints on span, width and thickness (see Chapter 3) are respected, because machine alignment and specimen/machine interaction are less important for flexure than for tension. Coefficients of variation of 0.02 or 0.03 are easily attainable, but the higher precision available by flexure must not be equated with higher accuracy. Most plastics processing routes produce samples and test pieces taken from them in which the molecular state and hence the mechanical properties vary through the thickness. Therefore, a test in flexure provides a beam stiffness datum but only a notional material modulus because each layer in a flexed beam contributes to the overall bending resistance in proportion to the square of its distance from the neutral axis. The elementary tables of properties published by plastics producers often include both tensile and 'flexural' modulus, the values of which are usually significantly different, reflecting the layered structure, the associated skin-bias effects and the need for characterizing properties data to be matched to the downstream patterns of utilization.

Modulus measurements in uniaxial compression are prone to variability and inaccuracy because a strut is potentially unstable under end-loads and the physical constraints and limitations on dimensions that are necessary to avoid buckling are detrimental to precision and accuracy. Modulus in uniaxial compression is seldom measured by ramp excitation.

4.9

If the test material is in its linear viscoelastic state, a modulus derived at constant deformation rate is related to the complex modulus and the relaxation modulus (Chapters 5 and 6, respectively) by integral equations, because the ramp excitation of the former situation can be equated to a succession of step excitations by application of the linear superposition principle. If the material is in its non-linear viscoelastic state, the relationships between the different response functions are not as straightforward as they are in the linear case but they retain a general similarity to them. Thus, though the response to a ramp excitation, i.e. a traditional force vs. deformation or a stress vs. strain curve, is questionable in that the effects of increasing strain and elapsed time are superposed, and though the isochronous test, see Chapter 6, is superior in that there is a clear separation of those effects, the former has a justifiable role within the framework of mechanical tests for viscoelastic materials. Their overall simplicity ensures their continuing popularity and the supporting rationale guarantees useful data provided only that the test conditions and data derivation procedures are specified and documented. Apart from that, ramp-excitation actuators driving specially devised subassemblies enable certain classes of downstream problem to be resolved expeditiously irrespective of the suspect accuracy of some results.

4.10

The characteristic of the excitation function must accommodate the viscoelastic nature of plastics in order that a property measurement yields more than a merely arbitrary value. The format adopted for these notes, namely the mechanically important properties modulus, strength and ductility being separately discussed in terms of specific excitation functions and the distinctive responses to them, was so chosen because of the dominant influences and sometimes insidious consequences of the viscoelasticity. However, that choice of format may seem to introduce an element of artificiality in the particular case of ramp excitation experiments. This is because the ramp, unique among the excitations, can be used (admittedly with some difficulty) to measure modulus, ductility and strength in a single test and consequently the measurement topics could be grouped together. Because of that practical feature and in the light of the format that has been adopted, this chapter and Chapter 9 have common threads of rationalization that are absent, say, in measurements of complex modulus and fatigue by means of sinusoidal excitation. The essential unity of classes of ramp excitation test is inherent in a paper by Roberts,[2] though that author's primary purpose was to model tensile stress–strain curves and derive 'reference points' therefrom rather than to display the unity.

References

1 I. T. Barrie, D. R. Moore and S. Turner. 'Developments in the generation and manipulation of mechanical properties data.' *Plastics and Rubber: Processing and Applications*, 3, 4, 365, 1983.
2 J. Roberts. 'Stress–strain property reference points for non-linear materials.' *Plastics and Rubber: Processing and Applications.*, 9, 1, 17, 1988.

S4.3   Sources of error in ramp excitation tests

There are several sources of error that affect the precision and the accuracy of modulus data generated in ramp-excitation experiments.[1] Those errors are variously attributable to:

1 practical limitations and imperfections in the coupling mechanisms for force transfer between actuator and test piece;
2 disparities between the desired (theoretical) stress field in the gauge section and the attainable field;
3 disparities between the notional excitation (a constant rate of straining) and the attainable one (a nearly constant rate of deformation after transients have disappeared);
4 inadequacies in the devices used to measure the deformation and the force.

To varying degrees, those sources of error are common to all mechanical tests and they are all interdependent. Thus, for instance, if the coupling between actuator and test piece is weak the test may function imperfectly or fail completely (the test piece may slip from the grips) and if it is too robust the stress field in the gauge region may be distorted to an unacceptable degree. The third source of error listed is particularly important in tests on plastics. Disparities between the desired excitation and the attainable one arise mainly from mechanical inertias apart from the practical reality that the excitation can at best be only a constant rate of deformation rather than a constant rate of straining. The mass of the actuator cannot be accelerated instantaneously to a constant velocity and therefore its initial movement occurs at varying speed as it accelerates from rest against an increasing resistance from the specimen. Hydraulically-driven actuators, which are 'soft' machines, may never deliver the deformation rate that is demanded if the deformation rate is high but with mechanically-driven actuators a constant-rate steady state is reached ultimately. However, the early, low-strain region of the response curve deviates from the true response, particularly at the higher deformation rates.[2]

Additionally, the progressive take-up of slack in the load transfer linkage contributes a false component to the apparent response. Thus there are uncertainties about the straining rate during the early part of the observed response and about the zero reference point for the derivation of a secant modulus.

The overall effect of those mechanical factors, apart from a superposed vibration if the deformation rate is high, is that the apparent force–deformation curve is initially concave upwards. A notional zero point can be derived by numerical curve fitting but that procedure normally encompasses the initial concavity and hence incorporates an error, though it is possible to extrapolate back from the region where the curvature is convex upwards and where, presumably, the initial transients have decayed to zero. The zeroing error or the uncertainty affects the accuracy with which a modulus datum can be extracted. Obviously, tangent modulus at the origin and secant modulus at small strains are likely to be erroneous. A strain of 0.002 for the latter would be in harmony with the common practice for isochronous creep modulus (see Chapter 6) but the modulus value could be unreliable; a strain of 0.01 raises the likely accuracy but intrudes into the region of viscoelastic non-linearity and therefore 0.005 is a compromise that is commonly adopted.

This type of test, particularly in the tensile mode of deformation, is widely used in support of product development, quality control, etc. for which it is eminently suitable by virtue of its ease and rapidity of operation. Another advantage is that, with a straightforward change of procedure, the same excitation and test configurations can be used to measure strength and ductility. Its limited accuracy when compared with that of step excitation methods (see Chapter 6) has not restricted its use for the purposes cited above and it seems unlikely that serious misinformation can arise provided ongoing data are utilized within a framework of corresponding archive data and quantified by links to other classes of modulus data. On the other hand, trains of inference based solely on ramp-excitation tests and with no independent corroboration could be fallacious in some circumstances.

References

1 S. Turner. 'Tensile testing of plastics.' Chapter 5, '*Tensile Testing*.' Ed P. Han, ASM International (1992), ISBN: 0-87170-440-4.
2 A. Kobayashi and N. Ohtani. 'Effects of initial non-constant transient cross-head speed on the stress–strain curve of viscoelastic material.' *Journal of Applied Polymer Science*. 16, 10, 2523, 1972.

S4.6 Tensile modulus measurements on continuous fibre reinforced composite specimens

The quality of tensile modulus measurements on heterogeneous, anisotropic materials, such as continuous fibre reinforced composites, is affected by the same factors as measurements on ordinary plastics but with some qualifications. Rate of straining is less important, whilst other factors are more important and these depend on the dimensions of the test pieces and arrangements for mounting them in the machines. In particular, the ends of test pieces should be reinforced by end-tabs to ensure non-distorting load transfer from the actuator and the dimensions of test pieces should be much greater than the size of any structural inhomogeneity or processing defect such as a kinked fibre or a void. The use of reinforcing end-tabs is currently almost universal but there is no general agreement about their size, shape and character; for example, one authority stipulates a minimum length of 38 mm and another stipulates a minimum of 50 mm. The details of the various recommended procedures have been decided mainly by *ad hoc* local experimentation, though the primary and common purpose is to limit the probability of inconveniently early failure initiating at a grip by local reinforcement of the material and reduction of the stress concentration at that critical point. End-tabs can also facilitate accurate alignment of the test piece in the test machine but, on the other hand, if they are fitted unsymmetrically they can cause misalignment and introduce stress concentrations.

Various test piece sizes and geometries are recommended, depending on the type of composite, i.e. the constituents, volume fraction and spatial distribution of the fibres. Thus, for example, one group of specifications postulate a certain test configuration for unidirectional laminates and a different one for multidirectional ones, whereas another group postulates a total of four configurations. Four standardization bodies postulate total lengths of 283 mm, 250 mm, 229 mm and 200 mm, respectively, corresponding free lengths of 175 mm, 150 mm, 127 mm and 100–150 mm and various widths and thicknesses.

If the tension axis does not coincide with the fibre alignment in a unidirectional composite, then the dimensions have to be carefully selected for the underlying reason that a 'representative number' of fibres should run from one side of the specimen to the other, clear of the end-tabbed region. One working guideline on that is the length should be greater than $w(1 + 1/\tan\theta)$, where w is the width and θ is the angle between the fibres and the longitudinal axis.

The various specifications are similar in their attention to details though they differ in the relative importance accorded to particular features such as the shape and character of the end-tabs, etc. Evolutionary changes to the specifications stipulated by the standardization committees reflect the progressive refinement of the collective understanding of the interplay of the many influential factors and also an increasing commercial need to reduce testing costs.

Little has been done to harmonize the specifications for modulus tests on continuous-fibre composites. A few, very limited, investigations in that area have been reported but the possible disparities in modulus values arising from the different dimensions have not been quantified nor has there been much public advocacy of the merits of international standardization. Active groups may have a vested interest in perpetuating the use of whatever procedures they adopted initially, because that enables them to make the best current use of their archive data, which are often extensive and which define modulus levels that have been shown to correlate with satisfactory performance. Additionally, composites are structures rather than materials and, correspondingly, the composites industry is mainly subdivided into sectors associated with specific classes of engineering applications, for which the standard tests may have to be narrowly and specially defined to be useful. Possibly as a consequence, little effort has been expended on the acquisition of facts about the relative merits of the various standard procedures.

In the absence of any direct evidence to the contrary one must assume that the various specified test procedures would yield different values for modulus (and strength) on samples of the same composition. Some slight light was shed on the issue in a paper by Sottos et al.[1] who compared the results obtained on one fibre–resin formulation in four different laminate lay-ups by use of three standard test methods with permitted variants. There were 17 sets of data, with five specimens in each set. The coefficients of variation for the modulus measurements varied from 0.013 to 0.120 with a mean value of 0.068. There was no indication that any one test method gave results with less variability than the others and therefore the mean values of the coefficients of variation may be taken as typical for this type of measurement. Thus, on the basis of a normalized distribution one might expect 95% confidence limits in the region of 0.08 to 0.10 for sets of five specimens, which is the recommended size even though five is the bare minimum set size for specimens exhibiting that level of variability.

A high coefficient of variation may be an inherent characteristic of fibre composite materials in general; for example, values of 0.079, 0.080, 0.083 and 0.100 for the tensile modulus of a polyester resin with four different volume fractions of glass fibre were tabulated by Sims et al.[2] If it is an inherent characteristic and if, because of that, the various standard tests do not give demonstrably different results, there is no reason for them all to be retained though, by a parallel argument, it should not matter if they are retained because the disparities could be largely ignored. Substantial savings in evaluation costs could be achieved if this matter were to be resolved.

A unidirectional carbon fibre composite with a fibre volume fraction of 0.60 might have a tensile modulus of about 120 GPa and even higher values are reported from time to time, which suggests that the fibre modulus is utilized with an efficiency of about 85%. However, modulus values vary with the type of fibre, the perfection of the alignment and other aspects of the structure.

Where the tension axis does not lie along a principal direction of the test coupon (if such a principal axis can be defined) the results may give information about the shear modulus.

References

1 N. R. Sottos, J. M. Hodgkinson and F. L. Matthews. 'A practical comparison of standard test methods using carbon fibre reinforced epoxy.' Poster 55, Sixth International Conference on Composite Materials, Imperial College, London, July 20–24, 1987.
2 G. D. Sims, A. F. Johnson and R. D. Hill. 'Mechanical and structural properties of a GRP pultruded section.' *Composite Structures*, 8, 1987, pp. 173–187.

S4.7 Unorthodox test configurations for the measurement of modulus

S4.7.1 In-plane flexure of a ring

One test configuration for in-plane ring flexure is shown in Fig. S4.7.1/1. In this particular sub-assembly, described by Whale and Stephenson,[1] the support rods are 10 mm in diameter and can be moved to give various spans. The actuator moves a central loading bar downwards and the bending resistance is derived from the force–deformation curve. For a circular ring the reaction at support lies along a radius of the ring and Roark[2] provides an elastic equation from which a modulus can be calculated, thus:

$$E = \left(\frac{P}{\delta}\right)\frac{r^3}{I} f(\theta) \qquad \text{[S4.7.1.1]}$$

where E = Young's modulus
      P = applied force
      δ = the deflection at the loading point
      r = the radius of the hoop taken at the centroid of the cross-section
      I = the moment of inertia
      $f(\theta)$ is given by the equation:

$$f(\theta) = \frac{1}{\pi}\left[(1+c)(1+c+\theta s) + \frac{\theta s^3}{3}\right] - s^3 - \frac{sc^2}{2} - \frac{(\pi-\theta)c}{2} \qquad \text{[S4.7.1.2]}$$

in which $c = \cos\theta$
          $s = \sin\theta$
See insert in Fig. S4.7.1/1.

Given the nature of $f(\theta)$, one would not expect a modulus datum calculated in this way to be highly accurate or precise, but there is technical interest in the variation of modulus around the circumference of pipes and pipe-coupling units. The variation arises from features of the fabrication process and can be importantly large. This ring test is useful because it can be carried out at different points around the circumference of a particular ring, thereby eliminating inter-ring variability, provided only that the deformation each time is limited to strains at which the ring suffers no damage or prolonged viscoelastic memory effect. Under such conditions the test geometry stays constant apart from variations in the moment of inertia due to thickness differences, for which allowance can be made, so that variations in the modulus can be measured relatively accurately. $f(\theta)$ need not be known if relative values are all that an investigation aims for.

Whale and Stephenson investigated a particular complicated coupling unit, injection moulded in glass fibre reinforced polypropylene through

S4.7.1/1. In-plane ring flexure test. If the modulus or the stiffness is deemed likely to vary around the circumference, the ring can be rotated relative to the supports and the test repeated, provided only that the maximum deflection in each test is small. The inset relates to Eq. S4.7.1.1.

four gates. The variation around the circumference of the quantity $\dfrac{[x-\bar{x}]}{x}$, where $x = \dfrac{P}{\delta h}$, and h = thickness, was ± 0.28; the peaks occurred where the flow geometry was such as to give favourable and unfavourable fibre alignments, respectively, in the circumferential direction.

Such large variations in the bending stiffness of what was a necessarily complicated device might affect its service performance under external loads and the thermal expansion coefficient would probably be similarly variable, with corresponding implications for the serviceability of the unit.

Thus, even though modern finite element analysis can provide estimates of the flow regime and the associated stiffnesses, experimental verification is necessary, at least in complicated cases such as this one, and therefore such subassemblies used in ramp-excitation tests have a useful role in product development programmes.

There are other ring test configurations for which there are analytical solutions using different forms for $f(\theta)$. A conveniently simple one is opposing forces acting outwards along a diameter, i.e. a ring in diametral tension. In that case, if point loading is assumed and deformations are small:

$$f(\theta) = \left[\frac{\pi}{4} - \frac{2}{\pi}\right] \approx 0.149 \qquad\qquad\qquad \text{[S4.7.1.3]}$$

Repeated testing along various diameters may be less informative than use of the configuration of Fig. S4.7.1/1 because the diametral configuration is less efficient at isolating particular sectors of the ring.

References

1 M. Whale and R. C. Stephenson. Private communication.
2 R. J. Roark and W. C. Young. *Formulas for Stress and Strain.* McGraw-Hill, Kogakusha, 5th Edn, 1975.

S4.7.2 Young's modulus in uniaxial compression and bulk modulus from one experiment

The bulk modulus and/or the bulk compliance of plastics are of great interest. This is mainly because of the light they may shed on polymer architecture, free volume and because of their relevance to three-dimensional stress fields in viscoelastic materials. However, they are rarely measured because of the exacting demands such measurements make on both apparatus and test procedure. The apparatus has to be elaborate and the results are always questionable because the confining fluids may penetrate the material, thereby contributing erroneously to the apparent volume change, or they may attack it. Thus, a simple device that yields an approximate answer might be regarded as a helpful alternative and Warfield proposed one such many years ago[1] that uses a simple subassembly in a conventional testing machine.

The apparatus consists of a steel cylinder and a plunger, which is driven down inside the cylinder by the actuator. A cylindrical test piece of slightly smaller diameter than the bore of the constraining cylinder is subjected to uniaxial compression by the plunger during the initial stage of the test, until its lateral expansion is prevented, at which point the stress field changes and further advance by the plunger creates a decrease of the volume. The force displacement curve shows a transition from a slope corresponding to the axial modulus (Young's modulus in uniaxial compression) to one corresponding to an approximation to the bulk modulus. Since the stress field is not truly triaxial the quantity measured is actually K_a, where, using elastic nomenclature for simplicity,

$$K_a = 3K \frac{1-v}{1+v}$$ [S4.7.2.1]

where K = Bulk modulus
and v = Poisson's ratio

Since v is not accurately measurable and since anisotropy in a test piece affects the issue, derived values of K are usually approximations but they can nevertheless be useful when cross-referenced to other modulus data. The test procedure also offers the operational advantage of two independent property values from one procedure.

Reference

1 R. W. Warfield, J. E. Cuevas and F. R. Barnet. 'Single specimen determination of Young's and Bulk Modulus.' *Journal of Applied Polymer Science*. 12, 1147, 1968.

Modulus from sinusoidal excitation tests

5.1

In the late 1940s and early 1950s the measurement of modulus by methods entailing test pieces being set into a vibration mode became popular, mainly because they offered a means by which frequency dependence and temperature dependence of the moduli of a viscoelastic material could be measured relatively rapidly. They were commonly referred to as 'dynamic mechanical tests'. It follows from Eq. S3.3.1 of Supplement S3.3 that the strain response of a linear viscoelastic material to a stress that varies sinusoidally with time is also a sinusoid. The response is not in phase with the excitation and therefore the ratio of the stress to the strain, i.e. the modulus, is a complex quantity (M^*):

$$M^* = M' + jM'' \qquad [5.1.1]$$

The ratio of strain to stress, the compliance C^* can be similarly described:

$$C^* = C' - jC'' \qquad [5.1.2]$$

$$M^* = \frac{1}{C^*} \qquad [5.1.3]$$

The phase angle between the input and response (δ) in a free vibration system can then be defined and is known as the loss:

$$\tan \delta = \frac{M''}{M'} = \frac{C''}{C'} \qquad [5.1.4]$$

In fixed vibration systems, which are commonly used now, the observed phase angle ϕ is not equal to δ.

If the material is non-linear viscoelastic, the response to a sinusoidal excitation is not truly sinusoidal.

5.2

The dynamic response of a Maxwell element, which is one of the two simplest mechanical analogues of viscoelasticity (a Hookean spring in series with a Newtonian dashpot, spring constant k, viscosity η) is that of a complex stiffness, k^*, where

$$k' = \frac{\omega^2 \tau^2 k}{1 + \omega^2 \tau^2} \qquad [5.2.1]$$

$$k'' = \frac{\omega \tau k}{1 + \omega^2 \tau^2} \qquad [5.2.2]$$

$$\tan \delta = \frac{1}{\omega t} \qquad [5.2.3]$$

$$\text{where } \tau = \eta/k \qquad [5.2.4]$$

The dynamic response of a Voigt element (a Hookean spring in parallel with a Newtonian dashpot) with the same constants is that of a complex compliance, c^*

$$c' = \frac{1}{k(1 + \omega^2 \tau^2)} \qquad [5.2.5]$$

$$c'' = \frac{\omega \tau}{k(1 + \omega^2 \tau^2)} \qquad [5.2.6]$$

$$\tan \delta = \omega \tau \qquad [5.2.7]$$

[Lower case c is used to preserve a distinction between the inverse stiffness of a Voigt element and the creep compliance of a material; it must not be confused with the lower case c traditionally used with appropriate suffixes to denote coefficients in the stiffness tensor of an anisotropic sample, see Chapter 7.]

The mechanical response of a single Maxwell or a Voigt element, corresponds only loosely to the observed behaviour of most plastics materials. Series combinations of many elements with different relaxation times, τ, have been widely used to provide a better correspondence, see Supplement S5.2.

5.3

M^* has the special virtue of being the viscoelastic counterpart of elastic M, in that all the elastic formulae are exactly correct for the linear viscoelastic case provided that M^* (ω) is substituted for M. Thus, whatever elastic solution is available for a vibrating test configuration is valid also for a viscoelastic test piece; for example, the elastic relationship:

$$K - \frac{GE}{9G - 3E} \qquad [5.3.1]$$

where K = bulk modulus

G = shear modulus

E = Young's modulus

becomes $K^*(\omega) = \dfrac{G^*(\omega)E^*(\omega)}{9G^*(\omega) - 3E^*(\omega)} \qquad [5.3.2]$

where $K^*(\omega)$ = the viscoelastic counterpart of K, etc.
for a linear viscoelastic material.

In contrast, for a creep or stress relaxation situation, the product term in the numerator is replaced by a convolution integral.

Other viscoelastic relationships for complex moduli are:

$$E' = 2G'(1 + v') - 2G''v'' \qquad [5.3.3]$$

$$E'' = 2G''(1 + v') + 2G'v'' \qquad [5.3.4]$$

and $\dfrac{G''}{G'} = \dfrac{E''}{E'}\left[1 + \dfrac{G'}{3K'}\left\{1 + \left(\dfrac{G''}{G'}\right)^2\right\}\right] \qquad [5.3.5]$

in which the value of $[1 + v'] \rightarrow 1.5$ for a rubber and $\rightarrow 1.25$ for a glass.

5.4

In principle, M^* may be measured by any procedure in which sinusoidal vibrations are imposed on a test piece and the amplitude and phase lag of the response are measured. The excitation is commonly imposed by a mechanical actuator and such procedures were earlier classed as 'dynamic mechanical' tests but acoustic (mainly ultrasonic) actuators, generally operating at much higher frequencies than the purely mechanical ones, are also used.[1,2] The latter tests differ fundamentally from the dynamic mechanical tests in that the wavelengths of the excitations are much smaller, and the enabling theory is that of wave propagation. The shorter wavelengths allow patterns of usage different from those for dynamic mechanical tests, for example detailed mapping of point-to-point variation in stiffness anisotropy in a moulded item and studies of heterogeneity. On the other hand, however, they introduce possibilities for operational difficulties if the excitation magnitude is similar to the size of some of the structural domains or to the lateral dimensions of the test piece.

The mechanical tests fall into two classes: forced vibration methods and free vibration methods, with the latter subdivided into non-resonance and resonance. In the former, the test specimen is overwhelmed by the apparatus and various frequencies are imposed over a range of temperature; in any one type of machine the range of available frequencies is not large, typically two decades of log frequency, because of mechanical and/or size limitations. In free vibration methods, the specimen is not dominated by the machine and therefore the operating frequency is governed by stiffness of the specimen so that as the temperature changes so does the frequency of the response.

Much effort was expended in the 1960s on the development of various types of test machine and test configuration, including vibrating plates, vibrating fibres and ball rebound. Later, other variants, such as the imposition of small sinusoidal excitations on test pieces already under steady static strain or stress, were used in studies of molecular structure and order, but such techniques tend to be used as research tools rather than as general purpose evaluation methods. (In contrast, that particular technique is widely used in fatigue/creep rupture studies because fluctuating stresses can often be severely detrimental to load-bearing lifetime.)

The literature abounds with papers describing the methods and presenting results but in retrospect most of the papers were of only transient interest because it transpired that many of the methods were impractical or inaccurate so that the options were effectively reduced to shear modulus from the free vibrations of a torsion pendulum (standardized for plastics in ISO R537, ASTM D2236 and DIN 53445) and tensile modulus from the free or forced flexural vibrations of a beam and from forced oscillating tension. Nowadays, furthermore, the availability of computer-controlled actuators and data processing facilities is biasing the choice towards forced vibrations. For rubbers, forced vibrations in shear are common; there are no severe practical difficulties and the results are especially pertinent to rubber items in service.

5.5

The accuracy of most of the dynamic mechanical methods is not as good as that attainable under step excitations (see Chapter 6) or even in certain ramp excitation tests. The factors contributing to the experimental errors vary with the test configuration and the method but arise mainly from inadequate compliance with St Venant's Principle and, in the case of composites, uncontrolled tension–shear coupling. The torsion pendulum is inherently more accurate than the vibrating reed because the relatively small dimensions of the latter magnify the importance of dimensional precision and distortions of the stress field at the loading points. Additionally, because of practical mechanical factors, such as available driving energy and mechanical inertias, the working strains tend to lie well within the linear viscoelastic region, which avoids analytical complications but adds to the potential for errors. Even so, there have been some investigations of complex modulus in the non-linear region, reference 3, for example.

Local distortion of the stress field and unintended vibration modes can arise if the test piece is heterogeneous and/or anisotropic, see Chapter 7. There are also attendant problems when the temperature range and/or the frequency range covered by the test is such that the modulus changes by a large factor (one or two decades is common in dynamic mechanical test programmes). This is because the interaction between machine and test specimen then changes during the course of the test. Reference 4 constitutes a good review of the methods and principles.

Poornoor and Seferis[5] studied the issue of accuracy from the standpoint of equating results from different types of test machine. They identified several sources of such differences in their experimental results and with appropriate corrective actions over the mechanical features of the tests and the different operating frequencies they were able to achieve nearly identical results from the different methods. They concluded that their study 'clearly showed that data from various instruments can be compared with one another as long as careful and proper data and instrument analyses are performed'.

5.6

Those practical limitations to the attainable accuracy tend to be outweighed for many users by the convenience and rapidity with which $M^*(\omega)$ can be measured as a function of temperature and frequency. It is commonly measured over a temperature range in a single (usually automated) operation, either at a constant frequency or at a frequency that varies with the modulus. The notional equivalence of frequency and temperature in their effects on complex modulus is only a rough guideline. The usual arrangement in such experiments is for the temperature to be raised at some constant rate; the recorded temperature may not correspond to the actual temperature of the specimen, partly because of temperature gradients in the apparatus and partly because of thermal inertia. Complex modulus should preferably be mapped as a function of both temperature and frequency.

Although the measured values of M', M'' and therefore also of $\tan\delta$ may be questionable the maxima in M'' and $\tan\delta$ occur at specific temperatures and frequencies governed by the molecular architecture, see Supplement S5.6 and Reference 6 and therein lies the main motivation for, and purpose of, dynamic mechanical testing.

5.7

The molecules vibrate randomly as a consequence of their thermal energy, but the movement of segments relative to neighbouring segments is hampered by potential energy barriers. The vibrational energy increases with temperature until certain classes of segmental movement are possible and mechanical excitations bias the potential energy field and thereby the direction of segmental motions, which is manifest macroscopically as a strain.

Under a periodic excitation the segments seek to resonate at their characteristic frequencies, but that response is inevitably out of phase with the excitation. The consequential imaginary part of the complex modulus is very sensitive to the vibrational state of the molecules and a graph of M'' or $\tan \delta$ vs. log frequency or temperature features well-defined maxima that locate and quantify the relaxation processes. M'' is a purer concept than $\tan \delta$ and it behaves well, even for a single Maxwell or Voigt element for which $\tan \delta$ does not. $\tan \delta$ values are unreliable also when M'' and M' both tend to zero. The real part of the complex modulus, on the other hand, like the relaxation modulus and the creep compliance, is a weighted summation of contributions from the various relaxation processes and is thereby not very discriminatory.

5.8

There was great interest in the mapping of complex modulus from the 1950s, due to demands for new polymers by the burgeoning plastics industry. Hundreds of different polymers and minor variants were characterized by the sinusoidal excitation techniques; reference 6, for instance, was one of the pioneering papers and, *in toto*, the many papers that followed constituted a major contribution to polymer science.[7,8] However, a surfeit of polymers and a degree of commercial stalemate subsequently dampened that architecture-related interest, though the technique continues to provide useful information and support in product development campaigns,[9] e.g. in the study of copolymers, blends, cure rates and cross-linking efficiencies in thermosetting polymers, etc. to the extent that special issues of journals are assigned to such topics[10] and dynamic mechanical analysis still has a respected place in mechanical testing strategy.

5.9

Complex modulus vs. temperature gives a comprehensive, but approximate, indication of some of the other mechanical properties of a polymer and, by inference, some indication of its potential for commercial use. Originally, its usefulness was overstated; it was quite common, for instance, for published values of M' to be used without modification in design calculations, despite wide disparities between the magnitudes of the strains and the durations of loading that prevail in the experiments and those commonly entailed in long-term load-bearing service. That misuse arose partly from a contemporary ignorance about the ramifications of viscoelasticity and partly from a paucity of other, more appropriate, data but there are now some approximate methods for using the dynamic mechanical test data to obtain more generally useful engineering moduli values and these are discussed in Supplement 5.9.

In contrast to the limited quantitative utility of M', the imaginary part of the modulus, M'', or more commonly $\tan \delta$, when presented as a function of temperature or frequency, is directly useful. Peaks on these plots denote regions in which mechanical properties change to such a degree that the functioning of a service item might be impaired.

5.10

The intensity of a loss process can be enhanced by the incorporation of a poorly bonded filler or by the use of blends of two or more polymers that have loss processes in adjacent regions of frequency and temperature. The resulting materials are used as coatings or in laminates to reduce resonances in metal sheets, see for instance another special issue of a journal.[11] Internal energy losses contribute very little directly to the attenuation of noise; surface friction between constrained layers is the major factor. Furthermore, the effectiveness of a damping layer is so sensitive to geometry of the assembly that complex modulus data for the material of the damping layer alone are insufficient for the assessment of fitness-for-purpose of the structure. Therefore, measurements of damping efficiency should preferably be carried out on the structure but otherwise on a strip of the laminate.

5.11

The utilization patterns for sinusoidal excitation tests have changed with the passage of time. In its initial phase, dynamic mechanical testing was a relatively primitive and painstaking operation that nevertheless became an essential tool in exploratory polymer chemistry and contributed outstandingly to the rationale linking mechanical behaviour to molecular architecture. Now, it is a largely automated routine operation that can still be used for that purpose but is often deployed at a more mundane level as follows:

- to ascertain quickly the general pattern of the relationship between $M*$ and temperature and excitation frequency;
- to check the effects of modifying additives on those relationships;
- to identify temperature and frequency regions in which $M*$ changes by a factor sufficiently large to be detrimental to service performance.

References

1 K. Thomas and D. E. Meyer. 'Ultrasonic measurement of reproducibility and anisotropy of processed polymers.' *Plastics and Rubber: Materials and Applications*, 1, 3/4, 136, 1976.

2 G. D. Dean. 'The application of acoustic surface waves to the characterisation of polymers.' *Plastics and Rubber: Processing and Applications*, 7, 2, 67, 1987.

3 M. N. Rahaman and J. Scanlan. 'Non-linear viscoelastic properties of solid polymers.' *Polymer*, 22, 5, 673, 1981.

4 G. D. Dean and B. E. Read. 'Measurement of dynamic viscoelastic properties.' *Plastics and Rubber: Materials and Applications*, 1, 3/4, 145, 1976.

5 K. Poornoor and J. C. Seferis. 'Instrument-independent dynamic mechanical analysis of polymeric systems.' *Polymer*, 32, 3, 445, 1991.

6 K. Schmeider and K. Wolf. 'Mechanical relaxation phenomena in high polymers and structure.' *Kolloid-Zeitschrift und Zeitschrift für Polymere*, 134, 149, 1953.

7 R. F. Boyer. 'Mechanical motions in amorophous and semi-crystalline polymers.' *Polymer*, 17, 996, 1976.

8 A. H. Willbourn. 'Molecular design of polymers.' *Polymer*, 17, 965, 1976.

9 M. J. Guest and J. H. Daly. 'Practical aspects of solid state mechanical spectroscopy for polymers.' *Polymer Yearbook*.

10 Eds. N. G. Mcrum and A. K. Dhingra. Special Issues *Plastics, Rubber and Composites: Processing and Applications*, 16, 4, 1991; 17, 5, 1992; 18, 3, 1992.

11 Eds. N. G. Mcrum and A. K. Dhingra. Special Issue *Plastics, Rubber and Composites: Processing and Applications*, 18, 1, 1992.

S5.2   The viscoelastic response of generalized Maxwell and Voigt elements

The response of a plastic to a mechanical excitation cannot be represented adequately by a single Maxwell element (spring and dashpot in series) or by a Voigt element (spring and dashpot in parallel). However, in some instances stress relaxation behaviour can be represented by a large set of Maxwell elements in parallel and creep behaviour by a large set of Voigt elements in series. Thus the concept of a wide distribution of elements with different relaxation times (τ), (retardation times in the case of creep) developed. The term k in the expression for k' in Eq. 5.2.1 had to be replaced by $f(\tau)d\tau$, with the dimensions of modulus, and the total response is the summation of the responses to each element.

The distributions of τ that are derived from experimental data are usually many decades of $\ln \tau$ wide and the summation procedure is simplified by the substitution $F(\ln \tau) = \tau f(\tau)$ and. The summation equations may then be written

$$M'(\omega) = M_o + \int_{-\infty}^{\infty} \frac{\omega^2 \tau^2}{(1+\omega^2\tau^2)} F(\ln \tau) d \ln \tau \qquad [S5.2.1]$$

$$M''(\omega) = \int_{-\infty}^{\infty} \frac{\omega\tau}{(1+\omega^2\tau^2)} F(\ln \tau) d \ln \tau \qquad [S5.2.2]$$

That alternative form was widely adopted because the distributions of τ that are derived from experimental data are usually many decades of $\ln \tau$

wide. In the literature $f(\tau)$ and $F(\ln \tau)$ are usually replaced by $g(\tau)$ and $H(\ln \tau)$, respectively, because the molecular relaxation processes occur mainly in association with the deviatoric part of the stress and strain tensors and the early experimentation on which the original mathematical modelling was based were correspondingly carried out in the shear mode of deformation.

The intensity functions $\dfrac{\omega^2\tau^2}{(1+\omega^2\tau^2)}$ and $\dfrac{\omega\tau}{(1+\omega^2\tau^2)}$ in Eqs. S5.2.1 and S5.2.2 effectively shorten the range of integration. In comparison with the normally wide spread of the distribution $F(\ln \tau)$ the former approximates to a unit step function at $\omega\tau = 1$ and the latter is small except in the neighbourhood of $\omega\tau = 1$ over which region $F(\ln \tau)$ may be regarded as approximately constant and therefore the integrals S5.2.1 and S5.2.2 can be simplified to:

$$M'(\omega) \approx M_o + \int_{1/\omega}^{\infty} F(\ln \tau)d\ln \tau \qquad [S5.2.3]$$

and
$$M''(\omega) \approx F(\ln 1/\omega)\int_0^{\infty} \frac{\omega\tau}{(1+\omega^2\tau^2)}d\ln \tau = \frac{\pi}{2}F(\ln 1/\omega) \qquad [S5.2.4]$$

which leads to the following useful approximations:

$$M'' \approx \frac{\pi}{2}\frac{dM'(\omega)}{d\ln\omega} \qquad [S5.2.5]$$

and
$$\tan \delta \approx \frac{\pi}{2}\frac{d\ln M'(\omega)}{d\ln\omega} \qquad [S5.2.6]$$

whence

$$\frac{2}{\pi}\int_{\omega_1}^{\omega_2} M''(\omega)d\ln\omega \approx \{M'(\omega_2) - M'(\omega_1)\} \qquad [S5.2.7]$$

and
$$\frac{2}{\pi}\int_{\omega_1}^{\omega_2} \tan\delta \;\; d\ln\omega \approx \ln\{M'(\omega_2)/M'(\omega_1)\} \qquad [S5.2.8]$$

Practical considerations generally dictate that M', M'' and $\tan \delta$ are measured primarily as functions of temperature and at only a few frequencies. Therefore, the calculation of M'' and $\tan \delta$ from M' via Eqs. S5.2.5 to S5.2.8 is prone to uncertainties and errors. The equations can be modified to forms that require differentiation with respect to temperature rather than $\ln(\omega)$, but the modifcation rests on further assumptions, with associated uncertainties and imprecisions, but see also Supplement S5.9.

Equations from S5.2.1 onwards form the basis of several methods for the derivation of $f(\tau)$ or $F(\ln \tau)$ from measurements of M' and M''. This was a topic of great interest at one time because, if $F(\ln \tau)$ is known from experiments using one type of excitation, the viscoelastic response to other types of excitation should be calculable. However, the inversion process is inherently imprecise and very sensitive to the accuracy of the experimental data which, as has been pointed out above, is not high for the majority of sinusoidal excitation experiments. In certain favoured cases, data for different excitation modes can be combined to widen the frequency range of the property map and other independent data, such as the activation energy for a particular loss process, may allow the M^*vs.T curve for one frequency to be adjusted to that for a different frequency. Although the overall accuracy of such manipulations is often limited, it can lead to a useful extension of the dynamic mechanical test data, as discussed in Supplement S5.9.

S5.6   Transitions in complex modulus

Since $M*$ is a complex number, the representation requires two quantities; the ideal choice is the real and imaginary parts, M' and M'', but M' and $\tan\delta$ are more usual. The positions of the $\tan\delta$ peaks are taken as defining 'transition temperatures'. Standard nomenclature for the transitions is α, β, γ etc., α relating to the highest temperature.

The locations and intensities of the peaks are governed by the molecular architecture; numerous correlations between transition temperatures and structural features have been established and published. Some peaks arise from segmental motion in the backbone molecular chain, some from segmental motions in side chains; some side chains function as main chains and some side chain stuctures influence the segmental motions in a main chain. Thus the rationale is quite complex.

In general, the peaks in $\tan\delta$ are sharp and more readily identifiable than the associated features on the M curves (vs. temperature or log frequency). Many amorphous plastics have a dramatically sharp α transition, but the high temperature flank can rarely be explored because the test piece tends to distort at that temperature and the response signal is either lost or heavily contaminated. The transition is much gentler in crystalline plastics and on the high temperature side the material is a crystal-reinforced rubber, the complex modulus of which can be measured; Fig. S5.6/1 is an example. Apart from the inherent complexities of the structure, an influential factor that was not clearly recognized during the pioneering studies of structure–property relationships is the effect of molecular order other than crystallinity. Thus, for example, differences in free volume, attributable to different rates of cooling through the rubber–glass transition and different residence times in critical temperature regions can affect the shape of the low temperature flank of that transition. Sometimes, this can occur to such an extent that the distortion takes the form of a peak. Figure S5.6/2 shows the effect of prolonged prior storage of PMMA at $100\,°C$. Such extreme differences in the state of order are not likely to feature in day-to-day experimentation but the example serves as a cautionary comment. It is possible that the observed shape of the low temperature flank of the main transition may differ from the 'true' shape because of the thermal conditioning to which the test piece has been inescapably subjected during the progression of the test up to that point.

Reports from various sources have also recorded similar effects attributable, for example, to the type of solvent used in the preparation of cast films and to the degree of coupling between the matrix and reinforcement in composites. Such results were usually accounted for in terms of a mathematical model appropriate for the situation.

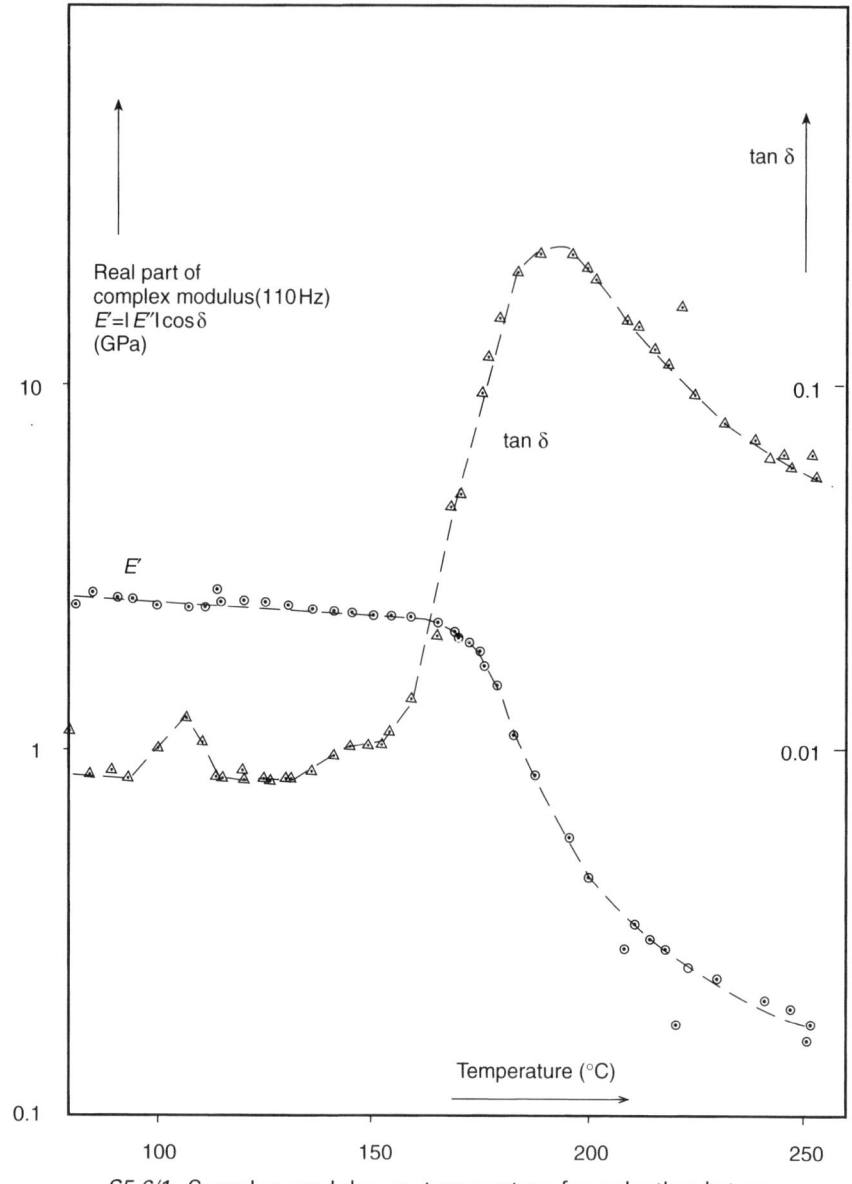

S5.6/1. Complex modulus vs. temperature for polyether ketone.
Above the main transition, the material is a crystal reinforced rubber.
(Unpublished information C. M. R. Dunn.)

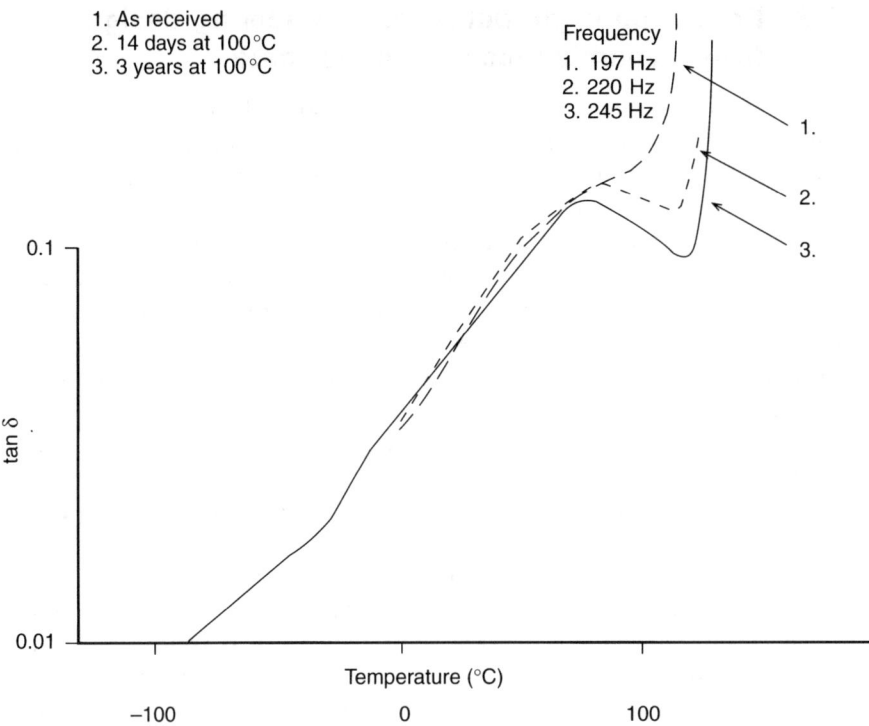

1. As received
2. 14 days at 100 °C
3. 3 years at 100 °C

Frequency
1. 197 Hz
2. 220 Hz
3. 245 Hz

1.
2.
3.

0.1

0.01

tan δ

Temperature (°C)

−100   0   100

S5.6/2. Tan δ for PMMA. The shape of the leading flank of the main transition is affected by the thermal history of the sample. (Unpublished information C. M. R. Dunn.)

The various effects are identical with those noted for other classes of modulus test and commented upon elsewhere in this monograph. It follows, taking account also of the relatively low accuracy of complex modulus measurements, that minor or secondary features on any of the response curves should not be accorded undue significance unless there is corroborative evidence or the phenomenon is the subject of the investigation.

S5.9   Prediction of modulus for engineering design from dynamic mechanical test data

There are two specific limitations in using moduli data from a dynamic mechanical (DMA) test for engineering design applications. First, the data lack accuracy in measurement. Second, the 'effective' times under load for these moduli are tiny compared with those that need to be accommodated in engineering application. There are additional limitations with regard to the level of deformation and the morphological nature of the samples, although such considerations are not peculiar to sinusoidal inputs and have in any case been discussed elsewhere.

The main limitations can be overcome by an approximate methodology that converts the test frequency to a time under load and then calibrates or adjusts the DMA data in the light of independent data derived by other methods in order to improve their accuracy and/or relevance to sustained load-bearing situations.[1] Figure S5.9/1 schematically demonstrates the principle of this procedure.

An experimental modulus versus temperature plot (DMA), at some specific oscillation frequency f is shown in Fig. S5.9/1. In addition, other frequencies were used in generating these data and loss factor ($\tan\delta$) was recorded during these experiments. Plots of $\tan\delta$ vs. temperature then revealed peaks which correspond to temperatures at which particular molecular transition processes occur in the material. An activation energy (ΔH) for this transition can then be obtained from an Arrhenius type relationship between temperature and frequency for the transition:

$$f = f_0 e^{-\left(\frac{\Delta H}{RT}\right)}$$

[S5.9.1]

where f_0 is a constant and R is the gas constant ($8.314\,\text{J/mol\,K}$).

If the transition is observed at two frequencies, the temperature of the transition will be T_1 for f_1 and T_2 for f_2. It then follows that:

$$\log_{10}\left(\frac{f_1}{f_2}\right) = \frac{\Delta H}{2.303R}\left(\frac{1}{T_2} - \frac{1}{T_1}\right)$$

[S5.9.2]

If the transition temperature (T_1) is known for a given frequency (f_1) then Eq. S5.9.2 permits the calculation of the transition temperature (T) for any other frequency (f):

$$T = \left[\frac{1}{T_1} + \frac{2.303R}{\Delta H}\log_{10}\left(\frac{f_1}{f}\right)\right]^{-1}$$

[S5.9.3]

For engineering load-bearing applications, it is more relevant to contemplate time under load rather than a frequency of oscillation. Therefore, the data are better fitted to a creep curve (with a step function for time) rather

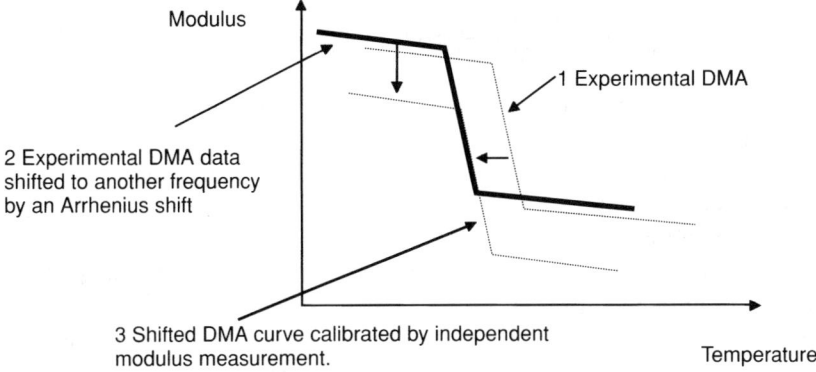

S5.9/1. Schematic for DMA calibration and shift.

than the sinusoidal input to DMA. An approximation due to Hamon[2] has shown equivalence between application of a constant load for t seconds to a frequency $(1/10t)$ Hz. Hence Eq. S5.9.3 becomes:

$$T = \left[\frac{1}{T_1} + \frac{2.303R}{\Delta H}\log_{10}(10f_1t)\right]^{-1} \qquad\qquad [S5.9.4]$$

This can be expressed in terms of a temperature shift $(\Delta T = T - T_1)$, which can then be used to shift the experimental DMA curve as illustrated in Fig. S5.9/1:

$$\Delta T = \frac{-2.303RT_1^2\log_{10}(10f_1t)}{\{\Delta H + 2.303RT_1\log_{10}(10f_1t)\}} \qquad\qquad [S5.9.5]$$

Assuming that the entire modulus-temperature function shifts by the same amount as the transition, then:

$$E_t(T) = Ef_1(T + \Delta T) \qquad\qquad [S5.9.6]$$

where E_t is a modulus at time t and temperature (T) and E_{f_1} is a modulus at the frequency f_1 and at the temperature $(T + \Delta T)$. In the region of the transition, this approximation is a good one. However, if other transitions are contributing to the observed modulus, then these probably have other activation energies and hence shift to a different degree.

The modulus values measured from DMA are generally of low accuracy. The limitations that occur from this can be avoided by calibration of the shifted DMA curve to a modulus value accurately determined in, say, a creep test (see Chapter 6). This is done by multiplying the measured DMA modulus at any temperature by the ratio of the creep modulus at 23 °C to the DMA modulus at 23 °C:

$$[E_t(T)]_{CAL} = \left(\frac{[E_t(23)]_{CREEP}}{[E_t(23)]_{DMA}} \right) \cdot [E_t(T)]_{DMA} \qquad [S5.9.7]$$

where subscript CAL denotes calibrated and subscript DMA denotes DMA measurements.

This second shift of the curve is illustrated in Fig. S5.9/1 and provides a modulus vs. temperature plot at the desired time under load and at an appropriate strain.

A list of activation energies is summarised in Table S5.9.1, where the values were obtained at three test frequencies in the range (0.1–10) Hz.

The accuracy of this method for converting standard DMA moduli values to a more practical modulus value at some specific temperature, and for a particular time under load, is limited and variable. In general terms, the nearer that temperature is to the transition temperature and the nearer the time under load to times equivalent to the DMA experiment, then the higher the accuracy. Additionally, peripheral evidence from other aspects of the mechanical performance often contribute to the overall view of the reliability of the procedure in specific cases.

Whatever its shortcoming, the power and value of the procedure, which to some degree supplements the convenience of the DMA technique with the higher accuracy of the creep experiment, should not be understated. Table S5.9.2 shows some data for a polypropylene copolymer, where a modulus predicted from this technique is compared with experimental creep moduli data for a range of times under load. The agreement between the measured and the predicted creep is not close, indicating that the attainable accuracy is far less than that typical of direct creep measurements, see Chapter 6, but the predicted time dependence seems to be accurate and, in this particular case at least, the modulus values are lower than those derived directly and therefore safe in a design context. In such operations, however,

Table S5.9.1 Activation energies for the α transition[1]

Material	Activation energy for α transition (kJ/mol)[a]
Polypropylene homopolymer	446
Propylene/ethylene copolymer (5% ethylene)	620
Polyether–ether–ketone	887
Polyethersulfone	969
Polystyrene	475
Acrylonitrile butadiene styrene	539

[a] (1 kJ/mol = 0.2388 kcal/mol).

Table S5.9.2 Comparison of measured with predicted modulus for a propylene–ethylene copolymer

Time under load (Seconds)	Experimental creep modulus (GPa)	Predicted modulus from DMA shift procedure (GPa)
10^2	0.58	0.53
10^4	0.39	0.34
10^5	0.31	0.26
10^6	0.24	0.188

one should keep in mind one of the concluding comments in reference 3, namely '... it may be concluded that this method can be used for design purposes, but that it calls for great caution and experience if gross errors are to be avoided. ...', even though it referred to a mathematical conversion method rather this semi-empirical one.

References

1 R. A. Chivers, M. Davies, D. R. Moore, J. Slater and J. M. Smith. 'Prediction of stiffness properties for design of engineering thermoplastics.' *Plastics and Rubber: Processing and Applications*, 14, 145, 1990.
2 B. V. Hamon. 'An approximate method for deducing dielectric loss factor from direct current measurement.' *Proceedings of the Institute of Electrical Engineers*, 99, 291, 1952.
3 A. Macchette and A. Pavan. 'Testing of viscoelastic function interconversion methods for use in engineering design. 1 Formulae to interconvert the transient functions (relation modulus and creep compliance) and the corresponding dynamic functions.' *Plastics, Rubber and Composites: Processing and Applications*, 15, 2, 103, 1991.

6

Modulus from step-function excitation tests

6.1

If a stress is applied suddenly and held constant thereafter, a consequential strain develops nearly instantaneously and then increases with the passage of time. The phenomenon is known as creep and the creep compliance $C_c(t)$ is defined as the strain arising divided by the applied stress. The relationship strain vs. log time is generally sigmoidal but the creep process is usually so protracted that the complete sigmoidal shape can rarely be inferred from a single experiment.

If a strain is applied suddenly and held constant thereafter, a consequential stress develops and then decreases with the passage of time. That phenomenon is known as stress relaxation and the so-called relaxation modulus, $M_r(t)$, is defined as the stress divided by the applied strain. Basic elements of linear viscoelasticity theory are given in Supplement S3.3, where it can be seen that the creep compliance is not strictly the elementary reciprocal of the relaxation modulus, see also Section 6.4, though the approximation is sufficiently accurate for most purposes. In most practical situations the behaviour is non-linear viscoelastic (see Supplements S3.3 and S3.5) though the definitions of creep compliance and relaxation modulus remain valid provided the applied stress (or strain) is quoted.

6.2

For thermoplastics in their glassy state, the rate of increase of strain under an applied stress, i.e. the creep rate, may be so low as to be judged zero by a casual observer. However, careful measurements on accurate apparatus always reveal a finite creep rate. These comments would apply, for example, to polystyrene and polyethersulphone at room temperature and for many untoughened thermoset materials. The usual method for presenting creep data for plastics is as strain vs. log (elapsed time). The main advantage of the logarithmic time scale compared with a linear scale is that it gives a conveniently open scale at both short and long times, both regions of which are important. The use of a linear time axis has been favoured for metals, mainly because they generally exhibit a section of constant strain rate, the region of so-called 'secondary creep', but any such region for a plastic is transient.

Log strain vs. log time is also a useful display because creep curves in that form often have extensive linear or near-linear regions, hence the frequent incidence of power laws in creep equations. The log–log plot is also useful when creep data are being manipulated numerically.

6.3

Creep and stress relaxation are important on two counts. Firstly, they are each the response to step excitations and thereby especially definitive within the framework of viscoelasticity theory and secondly, they relate to the load-bearing capability of end-products. Stress relaxation was favoured at one time as a supporting technique by polymer chemists seeking to relate the properties of polymers to their molecular architecture. Creep has been the favoured method for the generation of modulus data for design calculations, the primary variables being stress magnitude and temperature.

6.4

The two groups of phenomena are closely interrelated; the creep compliance function contains the same information as the relaxation modulus function, though it is expressed in slightly different form. For a linear viscoelastic material the relationship can be described through the convolution integral:

$$\int_0^t M_r C_c(t-u)\mathrm{d}u = t \qquad [6.4.1]$$

where $M_r(t)$ = relaxation modulus
and $C_c(t)$ = creep compliance

The relationship approximates to reciprocity when the time dependence is slight, see Chapter 3 Supplement S3.3.

In many situations the requirements for quantifying the modulus and compliance functions of Eq. 6.4.1 overwhelm the attainable level of accuracy, precision or discrimination of most test machines. In addition, at low values of elapsed time, the equation requires that $M_r(t)$ or $C_c(t)$ be known from time zero, which is not possible because mechanical inertias dictate that an exact step excitation is unattainable. However, the approximation of simple reciprocity often suffices, except where the properties and the boundary conditions are changing rapidly, e.g. during rapid loading (which is where the requisite data are inexact). Apart from that, Eq. 6.4.1 is seriously deficient if the material is non-linear viscoelastic.

6.5

In principle, creep and relaxation tests are simple but, in practice, neither class is straightforward if high accuracy is required. In the first instance, it is important that the stress field or the strain field should be a close approximation to that intended and that the measurement system should be adequately sensitive and accurate. However, tests of long duration entail additional considerations, namely:

1 the stability of the strain sensors over long periods of time;
2 changes in the molecular state of the test piece during a test, i.e. changes that will occur without application of stress;
3 dimensional inconstancy of the test piece due to minor variations in the ambient conditions.

For instance, temperature fluctuations of the equipment or the progressive uptake of water by the test piece, can give rise to spurious trends in the data. This is particularly so at long elapsed times and/or when the creep or relaxation rate is low. Extrapolation of the data to longer times may then be even less reliable than normal. Avoidance of such measurement pitfalls tends to require special features in machine design and operating procedures. (See Supplement S6.5.1 for some general comments on machine design for creep tests.) Also, details of standardized test procedures can be found in documents such as:

> ASTM D2990-77(1982) *Standard Test Methods for Tensile, Compressive and Flexural Creep and Creep Rupture of Plastics.*
> BS 4618: *Recommendations for the Presentation of Plastics Design Data*; Section 1.1 (1970) Creep; Subsection 1.1.1; Subsection 1.1.2; Subsection 1.1.3.
> ISO 899-1981. *Plastics – Determination of Creep Behaviour.* Part 1: *Tensile Creep* and *ISO/DIS 899-1.*

More recently, and reflecting the latest research, pertinent information is given in:

> *Code of Practice for the Measurement and Analysis of Creep in Plastics* by P.E. Tomlins, NPL Best Practice in Measurement Series, NPL MMS 002:1996, ISBN 0 946754 16 0.

6.6

Both types of experiment yield a time-dependent modulus, $M_r(t)$ from a relaxation experiment and $M_c(t)$, the elementary reciprocal of the creep compliance $C_c(t)$, from a creep experiment. The latter definition evolved in response to pragmatic demand from prospective users of creep data, but it is not entirely satisfactory in that the observed decrease with increasing elapsed time includes a strain-dependent component. In both cases, the viscoelastic effects are rationalized in terms of a wide distribution of relaxation (or retardation) times which, in turn, relate to the molecular architecture. Those distributions or the long-term course of stress relaxation can be derived approximately from stress relaxation master curves created by time–temperature superposition of short-duration experiments conducted at various elevated temperatures, see Supplement S6.6. In general, the technique cannot be used with any degree of confidence for creep data because the creep response usually includes a progressively increasing non-linearity contribution whereas, during a stress relaxation test, the non-linearity contribution remains constant at a level governed by the magnitude of the applied strain. Where necessary, creep data are extrapolated directly but cautiously; the procedure can be supported by notional time–temperature superposition in that short-duration creep tests at higher temperatures are useful indicators of likely trends in the actual creep response at a lower temperature during the period covered by an extrapolation. Creep data for a higher stress can be similarly used in support of an extrapolation.

6.7

The shear mode of deformation and the bulk mode have different time dependencies, as would be expected from the elementary model of molecular mobility. Mixed mode deformations reflect the differing time dependencies of their constituent components, which is particularly important in the case of fibre-reinforced composites because of the additional influence of the reinforcing phase, see Chapter 7.

The majority of creep and relaxation studies have been directed at the tensile mode and the shear mode. A shear configuration is the most sensitive to elapsed time, but a tensile configuration conforms with the traditional approach to engineering design for load-bearing service. The bulk mode has been almost totally neglected except to satisfy scientific curiosity but Benham and Mallon[1] compared creep under uniaxial, shear and hydrostatic conditions in PMMA, PVC and PP. The experimentally derived bulk creep modulus at 10^2 seconds, 10^4 seconds and 10^6 seconds was compared with the values predicted from the uniaxial and lateral contraction data on the assumption that:

$$B(\sigma, t) = 3C(\sigma, t)\{1 - 2v(\sigma, t)\} \qquad [6.7.1]$$

where B = bulk compliance
        C = uniaxial compliance
        v = lateral contraction ratio
        σ = tensile stress
        t = elapsed time

The agreement between the measured and the predicted values was good for the PMMA and the PVC but poor for the PP. The discrepancy in the PP data may have been due to stress-induced structural changes, to anisotropy in the test pieces or to Eq. 6.7.1 being time-dependent elastic in form rather than properly viscoelastic.

The creep lateral contraction ratio and the relaxation lateral contraction ratio are measured as almost essential supplements to tensile creep tests. One or other of the ratios is required for most engineering design calculations (though an approximate value often suffices), they shed useful light on volume change during creep and on the degree of anisotropy in the test pieces. Accurate experimentation is difficult, as was discussed in Chapter 4, but once the transients associated with the step loading have disappeared, the quieter conditions that prevail in creep and relaxation tests as compared with ramp excitation tests are more favourable to accuracy and also on the levels of anisotropy.

6.8

Numerous creep equations have been proposed to represent the strain response to an applied stress, mainly for tensile creep. Initially, many were of a form where the variables of stress and time were separated:

$$\varepsilon - f(\sigma)g(t) \qquad [6.8.1]$$

but later most equations accommodated the variables being inseparable and took the form:

$$\varepsilon = F(\sigma, t) \qquad [6.8.2]$$

Some include temperature as a variable. None have been successful except as models of particular sets of data. Their predictive value as a basis for extrapolation has a poor record.[2]

It transpired, long ago, that the non-linearity involves far more than the overt departure from linear proportionality between stress and strain by which it is usually defined. It disrupts the principle of superposition by which the response to a time-varying excitation can be derived from the response to one of the standard excitations. The effects may be interpreted loosely in terms of fading memory and there is strong supporting evidence from situations in which the stress is reduced as time passes; for example, recovery after creep of long duration is much tardier than it is after creep of short duration.

6.9

The most important indirect consequence of viscoelastic non-linearity is that recovery after creep, cannot be predicted accurately by linear super-position theory. Nor can it by any of the non-linear equations that have been postulated so far. Thus, recovery, which is an important phenomenon because it mitigates the creep strain and permits reduced load-bearing cross-sections when the service stresses are not sustained continuously, has to be regarded as a phenomenon in its own right. The same is true for other stress histories with the one important exception of regularly intermittent stressing which seems to excite a pattern of response that can be predicted on the basis of very simple assumptions. In general, however, the response to an arbitrary excitation history can rarely be deduced by calculation and consequently there is currently no alternative to extensive and meticulous experimentation and the approximate simulation of likely service situations.

The particular case of recovery is discussed in Supplement S6.9. Whereas for creep at one temperature the influential variables are applied stress and elapsed time, for recovery the influential variables are the elapsed recovery time, the duration of the preceding creep and the magnitude of the final creep strain. In addition, there are experimental difficulties peculiar to the recovery situation so the testing burden is very high if comprehensive mapping is attempted.

6.10

As a consequence of the widely prevalent non-linearity, the primary form for creep data has to be a family of creep curves at a range of applied stress levels. Figure 6.10/1 is an example. It includes quite a few creep curves at different levels of stress. However, fewer curves may suffice for certain thermoplastics in their glassy state, some thermosetting resins and continuous fibre composites.

Generally, nowadays, only some of the individual curves in that family are experimental results because it has transpired that many classes of plastics each exhibit a characteristic pattern of creep behaviour which, once identified, can be quantified by a limited number of creep tests. The details

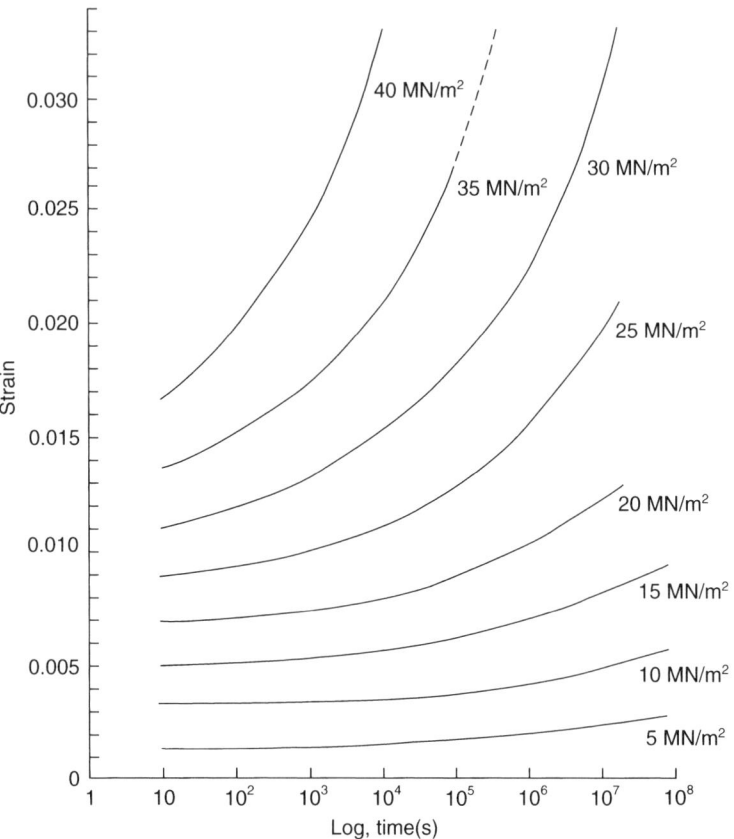

6.10/1. Tensile creep curves for a PVC at 20 °C. The experimental curves can be supplemented by notional curves for intermediate stresses by numerical interpolation if the general pattern of behaviour can be identified. Some experimental curves can be extrapolated to a limited degree if the curves for higher stresses do not suggest that an upward swing is imminent for the curves in question. Each class of plastic displays a characteristic pattern of creep behaviour that can be identified by an appropriate test programme and used to advantage in support of interpolation and extrapolation procedures.

vary from class to class but the overall evaluation process is simplified by the use of a particular experimental procedure, see Supplement S6.10, that yields an isochronous stress–strain relationship, which quantifies the non-linearity at a selected elapsed time and thereby provides a spacing rule for the creep curves.

It can be argued that the curves at the higher stresses in displays such as Fig. 6.10/1 relate more to the phenomena of creep rupture than to those of creep. This is because they presage failure, and if that is so the critical feature can be derived more economically by the measurement of lifetime, see Chapter 10. Even so, a sharp upward trend of a strain vs. log time curve is a more quantitative indicator of stress levels that should be avoided in service, of ductility and of aspects of failure other than lifetime data from creep rupture tests.

For downstream use, the creep curves are often transformed into modulus vs. log time. However, the creep modulus, $M_c(t)$, defined in Section 6.6, is not ideal and isometric creep modulus derived from sections at various constant strains across the family of creep curves is the preferred form for the data. For that purpose, sections are commonly taken at tensile strain intervals of 0.005 up to 0.030 for the polyolefines, but to lower upper limits for thermoplastics in their glassy state, for thermosetting resins and for polymer–fibre composites, the limit depending on the ductility and extensibility of the material.

6.11

Isometric modulus vs. time is only sufficient for the resolution of the most straightforward load-bearing problems and therefore a family of recovery curves, for various combinations of preceding creep strain and creep duration, are an important subsidiary set, see Fig. 6.11/1, for example. There are

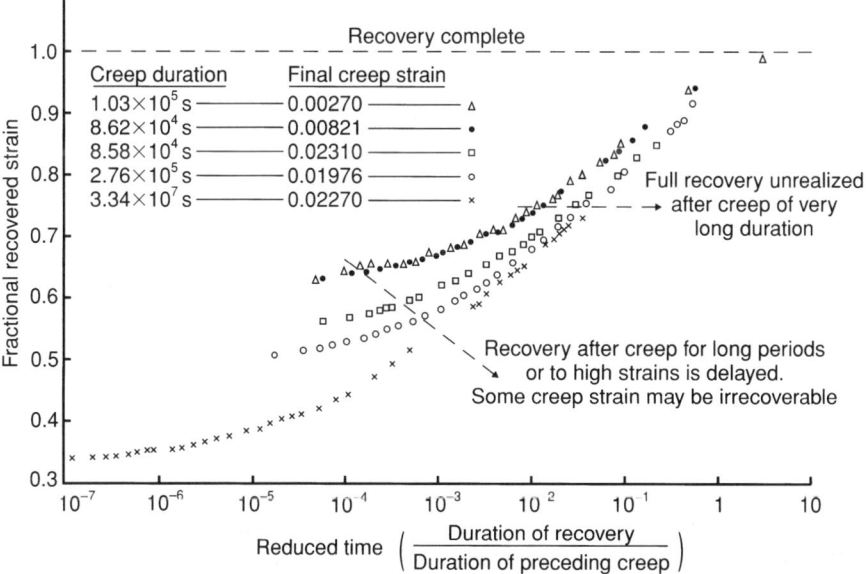

6.11/1. Typical recovery data – nominally dry acetal copolymer at 20 °C. Such data cannot be deduced by linear superposition of creep curves.

various ways in which recovery data can be displayed; the format adopted here and in Supplement S6.9 is unsuitable for computation via a superposition theory, but it conveniently condenses data and relates the course of recovery to the preceding creep, which is of direct practical importance. There often seems to be an irrecoverable component in the strain even when the creep is well below what would normally be regarded as the yield point, but see Supplement S6.9 for further comment. Furthermore, the apparent yield point on a conventional stress–strain curve, see Chapters 8 and 9, tends to accentuate what is often a gradual transition from a viscoelastic state to a plastic state. Therefore, it is feasible that genuine irrecoverable strain may develop ostensibly below the yield point, just as some of the strain developed beyond the yield point may be recoverable.

6.12

The strain responses to more complex stress histories consist of largely unquantified non-linear viscoelastic superposition complicated by changing properties due to progressive molecular rearrangements coupled with work hardening or work softening processes.[3] In general, therefore, they are not amenable at present to mathematical representation and analysis; for the current extension of the linear superposition theory to have any predictive value for arbitrary loading paths, the critical quantities would have to be measured to a precision ten times finer than that currently attainable. The experimental limitations arise from inherent mechanical constraints in creep apparatus and unavoidable interspecimen variabilities. The latter distort the small but important differences in the creep rates that occur in nominally identical test pieces. In turn, this leads to a lack of reliability in any mechanical model being derived. In view of those difficulties, interest in this subject has waned and has turned to the identification of characteristic patterns of behaviour from which, in conjunction with the primary and subsidiary (recovery) data sets, reasonable estimates of the response to some complex stress histories can be deduced.

6.13

The practically important case of regular intermittent stressing is one for which non-linearity causes no serious analytical difficulties. Experimental investigations in both straightforward intermittent stressing and of the isochronous stress–strain procedure have shown that the unrecovered, but ultimately recoverable, strain from successive creep periods tends to accumulate progressively, but that the pattern of behaviour is simple and regular, and the strain vs. accumulated intermittent creep–time relationship

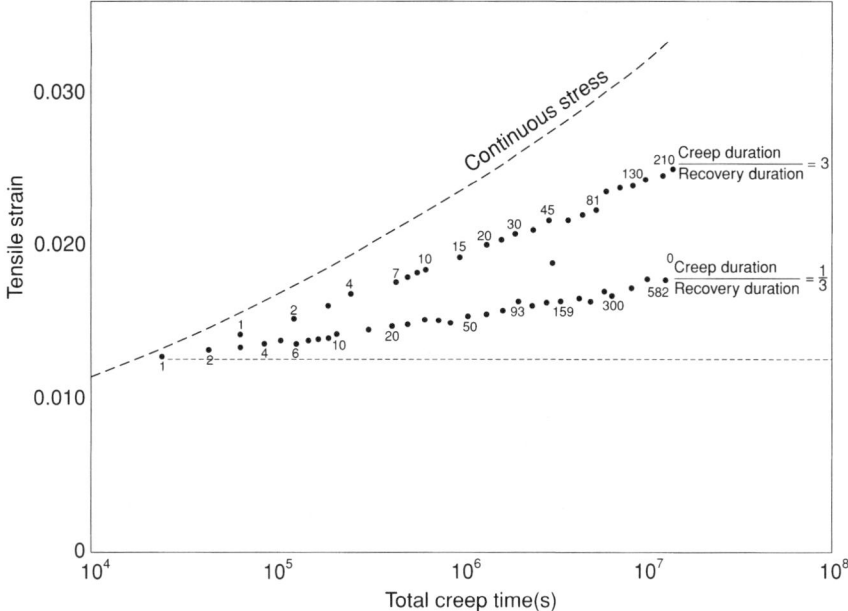

6.13/1. Creep under intermittent stress, diurnal cycles, polypropylene, 20 °C, 10 MPa. The numbered datum points record the final creep strain at the end of the nth cycles. The simple pattern of behaviour during successive cycles is found for other periodicities and with other plastics. The maximum creep strains attained in each cycle can be converted into effective modulus as a function of total time under stress.

(ignoring the intervening recovery periods) always falls below the strain vs. uninterrupted creep time relationship.

The response to intermittent stressing follows a remarkably simple pattern, see Fig. 6.13/1, despite the relative complexity of recovery behaviour.[3] A graph of maximum strain reached during the rth creep period vs. the logarithm of the cycle number, r, or vs. the logarithm of the total time under stress, is a nearly straight line which tends to curve upwards at high values of the abscissa, due to an accumulation of residual strain. The overall effect was found to be impressively regular and simple for polypropylene and for acetal copolymer at several stress levels and for different cycles. There was some evidence also that the same pattern arises with glassy amorphous polymers. An elementary analysis justifies a set of semi-empirical rules and the limiting values of the strain reached during or under intermittent stress divided into the applied stress give an 'effective modulus' (an $M_c(t)$ rather than an isometric modulus) which can be used in design calculations.

Two important points emerge

1   The response to intermittent stress follows a very simple pattern, even though the behaviour is actually complex and beset with secondary effects (which have not been mentioned here, but see reference 3);

2   The conversion of the data into 'effective modulus' reduces a seemingly complex situation to a form readily usable in design calculations.

Empirical rules such as those mentioned above are of limited generality. Their validity should not be assumed, particularly in regions which have not been explored experimentally. This is because the pattern of behaviour may vary locally or different classes of phenomena may intervene. For example, though it is likely that all plastics will deform in much the same way under intermittent stressing, some are prone to embrittlement or premature failure under such conditions. Thermoset materials and amorphous thermoplastics in their glassy state seem to be particularly susceptible and for them the experimental considerations are those of Chapter 10 and the crucial design consideration may be the creep rupture lifetime rather than excessive creep strain.

6.14

All mechanical test procedures have refinements designed to provide accuracy commensurate with the purposes of the test. However, such procedures within the broad class of step excitation tests provide certain data adjustment routes that are not possible or have not yet been developed for the ramp excitation (Chapter 4) and sinusoidal excitation (Chapter 5) classes. Thus, for example, the isochronous stress–strain data serve their own specific purpose. However, they also provide checks on the accuracy of the creep curves and other supporting evidence, whilst recovery data are necessary in certain design situations but also provide information about the quality of the preceding creep test and the nature of the creep strain.

Essential support for creep tests at elevated temperatures can be obtained also from ramp excitation and sinusoidal excitation tests at high temperatures. Ramp excitation tests can show the effect of temperature on the short-term modulus more rapidly, though less accurately, than the isochronous test can. Sinusoidal excitation tests provide similar information but also detect loss processes that may be indicators of poor long-term creep resistance at particular temperatures and thereby identify the temperature regions in which creep tests are warranted or essential. These supplementary procedures can identify critical regions and thereby possibly reduce the number of elevated temperature creep tests deemed to be necessary.

The overall process will generate large data sets, which poses certain handling and management problems. These include problems associated with the manipulation and storage of data, as well as decisions about what the format of a creep database should be and the choice of the logic paths governing a datum appropriate for a particular downstream problem from a large array of creep property data. Some simplification has to be effected at the data presentation stage, because of the nature of design calculations. It seems likely that data generators must play some part in that simplification process since they should be more aware than the designers of the ramifications of each simplification or omission. Similarly, in a reverse feedback process, data generators may have to react to evolving data usage patterns by modifying data presentation formats and possibly the enabling test procedures. Some of the issues are discussed in Supplement S6.14.6.

References

1 P. P. Benham and P. J. Mallon. 'Creep of thermoplastics under uniaxial, shear and hydrostatic stress conditions.' *Journal of Strain Analysis* 8, 277, 1973.
2 T. Sterrett and E. Miller. 'An evaluation of analytical expressions for the representation of plastics creep data. *Journal of Elastomers and Plastics* 20, October, 346, 1988.
3 S. Turner. 'The strain response of plastics to complex stress histories'. *Polymer Engineering and Science* 6, 306, 1966.

S6.5   Creep testing – apparatus and procedures

General principles

Creep test machines for plastics have been developed mainly as modifications of those used for the past century for metals, with differences to
accommodate the lower moduli of plastics. For instance, a small frictional
resistance that would be insignificant in a metals testing machine can introduce serious discrepancies in measurements of relatively soft, low-modulus
materials and a self-alignment mechanical system that relied on some resisting stiffness of the test piece would likewise be ineffectual. The availability
of suitable test pieces is another significant difference between the metals
and the plastics situations because typical section thicknesses of plastics
service items lie between 1 and 6 mm, which discourages the use of larger
test pieces, the properties of which are also likely to be atypical. In the light
of recent recognition of the influence of flow geometry on the properties
of thermoplastics the use of two thicknesses, e.g. 1 mm and 4 mm is now
advocated, the flow geometry usually being radically different in the two
cases. An international standard for thickness of 4 mm has been adopted in
many of the mechanical tests.

Apart from such mechanical factors, accurate creep testing of plastics
entails special provisions during the earliest stages of a test and during the
later stages (say after a few days). The need for the former arises as follows:

The loading path at the beginning of most creep tests on metals need not be very carefully controlled because the initial response is mainly elastic and the creep strain develops slowly. In contrast, many plastics creep during the loading process, because the retardation spectrum includes some very short response times and the identificaton of an 'elastic' component is virtually impossible. Effectively, the excitation is not a true step function and therefore the creep compliance function cannot be derived exactly at short times. As the elapsed time grows large in relation to the loading time, the disparity becomes insignificant. In addition, some discrepancy in the measured response at short times is tolerable when the main preoccupation is the response at long times, as it often is in creep studies, but for certain types of analysis the imprecisions are troublesome.

A good working rule, justified by both analysis and experiment, is that creep data for elapsed times less than about ten times the loading period should be treated with reserve. Where strain vs. log time curves appear to be convex upwards at short times the explanation is almost certain to be a lengthy loading period or ambitiously early strain readings. Clearly, the more abrupt the loading path the shorter the period over which the strain is erroneous or suspect and therefore most modern creep machines for plastics incorporate some device to achieve rapid but shock-free loading.

The source of error during the later stages of a creep test on a plastics material is a propensity for extraneous dimensional changes to occur. These changes may be variously due to:

1   the slow redistribution of the internal stresses set up during the moulding and other fabrication processes;
2   gradual equilibration with the environment;
3   property changes due either to progressive changes in the molecular state of the sample or to interaction with the environment.

In all cases, the rate of change is governed largely by the same molecular mobilities that control the creep rates, and hence they are likely to become manifest gradually during the course of creep or other long duration experiments.

It is recommended that the change of dimensions should be monitored on an unstressed control specimen during creep experiments, to enable the experimental curve to be corrected for at least some of the inadvertent dimensional changes. Changes in the molecular state are less easily handled. From a scientific viewpoint, the best procedure would be to confine the experiments to samples that are in a stable state. Special annealing procedures may be used to give an approximation to stability in crystalline plastics, but the mechanical properties are then usually atypical of the material as perceived in service. The situation is more complex for glassy amorphous plastics, especially at temperatures within about 30 °C of their glass–rubber

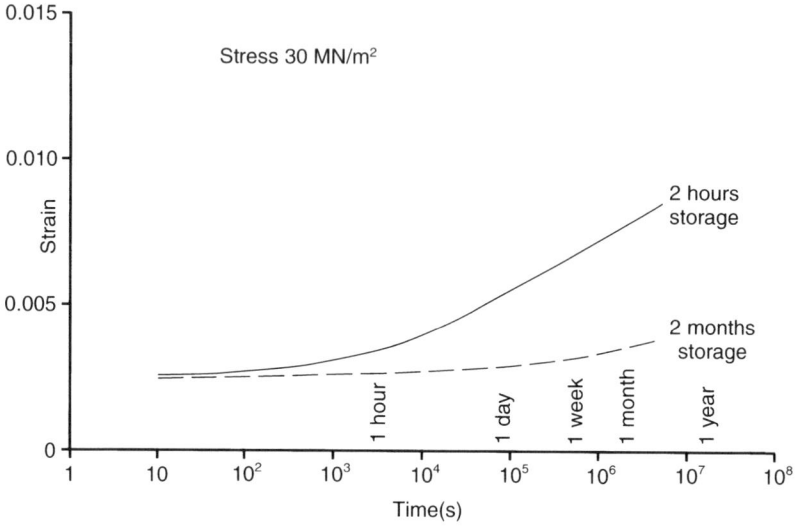

S6.5/1. The effect on tensile creep of preconditioning at the test temperature. Glass fibre reinforced polyethersulphone at 180 °C. The creep behaviour of all thermoplastics in their glassy state is sensitive to the thermal history of the test piece and particularly so at test temperatures near to the glass–rubber transition temperature. All such creep data should be qualified by full details of the storage history.

transition, which happens also to be the region that can yield critical creep data when load-bearing capability is being evaluated. There seems to be no dominating pretreatment equivalent to the annealing procedures available for crystalline plastics and therefore experimental creep curves for glassy amorphous plastics are not accurate responses to step excitations. The distortions are often very severe, see Fig. S6.5/1, for instance.[1] The experimenter cannot eliminate such distortion of the curves, because of the nature of the phenomenon; he/she can merely be aware of it and quantify data accordingly.

If the test programme is dominated by severely practical requirements, it may be argued that a sample in some 'average' state is the only appropriate one, though this requires that the particular state should be defined and maintained as closely as possible throughout the test. Consequently, the inevitable progressive change in the material state should be registered by independent tests. The properties of some polymers are sensitive to water, for instance. The rate of equilibration is governed by the rate of diffusion, which is often so slow that special conditioning procedures are resorted to. However, those procedures give only approximations to equilibrium and experimental results for samples so conditioned may be

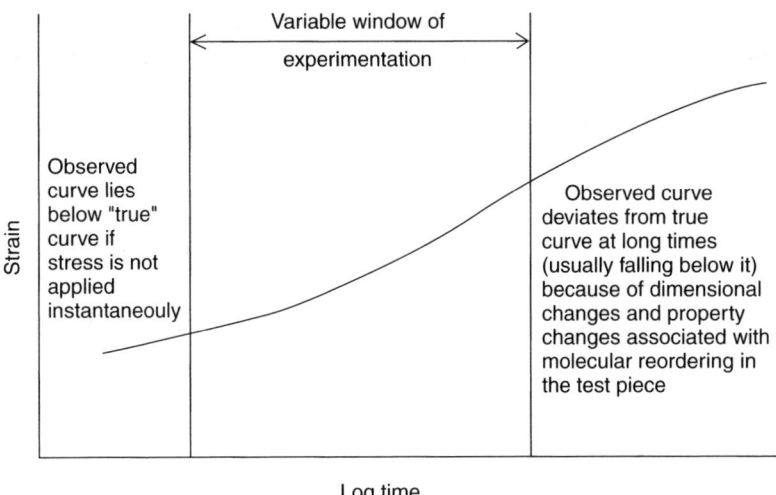

Variable window of experimentation

Strain

Observed curve lies below "true" curve if stress is not applied instantaneouly

Observed curve deviates from true curve at long times (usually falling below it) because of dimensional changes and property changes associated with molecular reordering in the test piece

Log time

S6.5/2. The window of experimentation and the disparities between the 'true' creep response and what can be observed.

actually less reliable than results for unrealistic simple states such as 'nominally wet' or 'nominally dry'. These issues arise for any type of long-duration test, including fatigue and creep rupture ('long' being defined relative to the timescale of the pertinent molecular mechanisms). This topic has received sporadic attention for many years but remains unresolved.

Because the step function excitation is inevitably imperfect and the internal state of the test piece changes slowly during a protracted test, an experimental creep curve is only an approximate picture of the creep behaviour. Thus, there is a 'window of experimentation', that varies with test temperature, material state, etc., outside of which the observed response may be unreliable, see Fig. S6.5/2.

Apparatus

Tension

The technical literature of the early 1960s abounds with descriptions of tensile creep equipment suitable variously for tests on plastics materials ranging from elastomers to fibre composites and in fabricated forms including fibres, films, laminates, etc. Tensile creep tests on fibres and films generally require only simple apparatus. However, more substantial test pieces require more elaborate test machines, the development of which was predominant, particularly those suitable for materials with a short-term modulus not less than about 1 GPa. This magnitude for modulus roughly marks the lower limit for materials likely to be of interest for load-bearing applications. The design of those machines followed the principles that had been proven for metals creep testing over the previous half century. However, they differed from country to country, small numbers of highly accurate machines being preferred in some geographic regions and larger numbers of machines of lower accuracy being preferred in other regions. Persuasive arguments can be advanced in favour of each option. Interestingly, nearly 40 years later regional differences in the prevailing testing strategies reflect the original differences.

The first modern tensile creep instrument designed specifically for plastics in the UK was due to Mills and Turner,[2] a version of which is shown in Fig. S6.5/3. It consists essentially of a reference face on the pedestal, to which is attached a low-friction guide block to ensure axial loading, and a subsidiary loading lever operating via a non-linear dashpot that can apply a load of 30 kg without shock within 0.04 seconds. The details are given in reference 1.

The strain is measured by an optical extensometer that is based on the Lamb and Cambridge instruments used for metals creep testing, with design differences necessitated by the low moduli of plastics relative to those of matals and their lower hardnesses. Also the sensitivity needs to be lower than that required for metals creep testing which is achieved by the use of large diameter rollers, though that entails a reduction of the mechanical stability.

The Mills–Turner machine was adopted by many other workers and modified by some. One such development is more compact than the original and yet accommodates the same specimen geometry. Axiality of stress is maintained by a cantilever spring connection between the movable hook and the pedestal. Instead of friction in a slide–guide system there is then a restoring force, in this case never more than 0.5% of the applied force. The extensometer consists of a frame bearing two diffraction gratings (250 lines per mm). Light transmitted by the grating pair forms Moiré fringes, which fall onto photodiodes. As the strain increases, the fringes move across the sensing element and their passage is counted, one fringe count corresponding to a strain of 2.5×10^{-5}, which is a sensitivity very similar to that of the Mills–Turner apparatus and Class A under Subsection 1.1.1 of BS 4618:1970. Another was designed to accommodate much smaller test pieces and has a similar displacement sensitivity achieved by use of differential capacitor transducers linked to it by extensometer arms attached through pivots to the main frame of the machine. The gauge is defined by the pointed tips of screws mounted on the extensometer arms.[3] The small interaction between this type of extensometer and the test piece makes it suitable also for the measurement of lateral strains though the extensometer attachments then become quite elaborate, often with counterbalanced arms.

S6.5/3. Tensile creep machine for plastics. This machine and several other models incorporate the design principles of the machines developed much earlier for creep tests on metals, but mechanical modifications were necessary to accommodate the lower moduli and hardnesses of plastics.

The effective gauge length for the lateral measurement is inevitably small which limits the sensitivity, repeatability and accuracy that is attainable. Evidence about attainable accuracy is scarce, but it is known that measured values of lateral contraction ratio are strongly dependent on the flow geometry, fibre alignment, etc. and that interspecimen variability is similarly sensitive and also much higher than that typical of tensile creep modulus measurements. The overall conclusion to be drawn is that one should be suspicious of any value of lateral contraction ratio that is quoted without accompanying identification of the test piece, its state of molecular alignment (and/or anisotropy) and the sample from which the test piece was taken.

Uniaxial compression

Compared with tension, uniaxial compression is a relatively neglected mode in creep studies. The few results that have been published show that thermoplastics are more resistant to creep in compression than to creep in tension. Therefore, tensile data, which are easier to acquire, provide a safe estimate of performance in uniaxial compression and that has sufficed except for limited special cases. The main difficulty encountered in uniaxial compression testing is a propensity of the test piece to buckle. That can be countered by lateral supports or by appropriate specimen geometry, e.g. relatively squat specimens, but both methods introduce an additional stress field, which stiffens the test piece. Jones et al.[4] adopted the former method and Thomas[5] used the latter. The apparatus developed by Findley employs a subassembly to convert the tensile force of his creep machine[6] into a compressive force acting along the axis of a long, slender test piece that is supported at intervals by spring-loaded rollers. That of Jones et al. is a similar device in which the support is provided by accurately fitting rings aligned in a tube. Thomas's apparatus relies on self-aligning thrust plates and other high precision features. The specimen is a cylinder of length up to 70 mm and diameter between 10 mm and 18 mm and the strain is measured by an optical extensometer with a gauge length of 30 mm. Thomas found that isochronous stress–strain curves (see Supplement S6.10) for test pieces of different lengths and with shaped ends were indistinguishable to within ±0.5%, from which she inferred that any errors in the measured strains due to barrelling were insignificant. The relatively large cross-section that confers stability on the test piece gives rise to a small error, corrected through the relationship:

$$\sigma = \frac{\sigma_a}{1 + \dfrac{\mu d}{3l}} \qquad\qquad [\text{S6.5.1}]$$

where

> σ = true stress
> σ_a = applied stress
> d = diameter of cylinder
> l = length of cylinder
> μ = coefficient of friction between specimen ends and thrust plate

The denominator probably has a value of about 1.01 under most conditions.

The equation applies to solid cylinders. A slightly different form is required for hollow ones, which are sometimes preferred. Thomas observed no difference between the creep responses of solid and hollow cylinders within the range of geometries that she specified but Mallon and Benham reported a size effect in tensile creep.[7]

Flexure

In some circumstances, flexure is a satisfactory alternative deformation mode to tension for the derivation of the tensile creep compliance function. In elementary experiments it offers the practical advantage that large, and hence easily measured, deflections can be achieved by the application of small forces. However, that advantage is often offset by the errors that accrue from possible friction at the supports and from discrepancies between the conditions of the experiment and the assumptions of the theory. The accuracy attainable in a casual test on simple apparatus is probably no better than ±10%. However, if the beam is long, thin and narrow, and if the deflection is no greater than one-half of the thickness, a measurement of modulus may be correct to within ±0.5%, which is better than could be achieved in tension at the same strain (0.0005).

There are three practical arrangements: the cantilever beam, three-point loading and four-point loading. The first two provide a measurement of the time dependent Young's modulus, or more correctly the reciprocal of the tensile creep compliance function, through Eq. S3.2.2.1 of Chapter 3.

The relevant equation for the third case is of the same type but slightly more cumbersome. In all cases the span should be large relative to the thickness to ensure that the shear strain is negligible and the deflection should not be large, to ensure that the applied force and the reactions at the supports are parallel to one another. Departures from the ideal can be countered in a purely geometrical sense by the use of more elaborate formulae, but a configuration that entails a large skin strain introduces viscoelastic non-linearity which cannot be accommodated in that way. In such cases the results should be expressed as deflections produced by particular bending moments rather than in terms of the creep modulus or its reciprocal.

The four-point loading method is the best because the beam is subjected to pure bending and therefore has a constant skin strain along the section between the two inner loading points. The skin strain is not constant along the span under three-point loading or for the cantilever; but that is unimportant when the test configuration is such that the material is in its linear viscoelastic state, which is approximately so for small deflection in long, slender beams. The cantilever method suffers from the complication that the physical situation at the clamped end of a beam differs from the mathematical form postulated in the Euler theory. In any case the end condition is ill-defined and variable between secure clamping which distorts the cross-section and insecure clamping which allows relative movement (equivalent to an unquantified increase in the effective length of the beam). Some results obtained by Dunn and Turner[8] and shown in Table S6.5/1 demonstrate the effect; though without independent evidence one should not assume that the result for the centre-loaded beam was more 'correct' than that for the cantilever with its end cemented into a metal block. The paper describes an apparatus for three-point loading capable of measuring the modulus to an accuracy of better than ±1%, discusses the more important sources of error and demonstrates that such accuracy entails painstaking attention to small details and the sacrifice of convenience in operation.

In elementary experiments, the values of width and thickness substituted into the beam equation S3.2.2.1 are averages, but when these quantities, particularly the thickness, vary along the length of the beam, the calculated result can be erroneous, and superior practice uses the modified equation:

$$E_c(t) = \frac{P}{\delta(t)} \sum_{i=1}^{n} \frac{A_i}{b_b h_i^3} \qquad [S6.5.2]$$

where the suffix i denotes specific values and the A_i are associated coefficients. Ten measurements of b and h made at equal intervals along the span seem to be sufficient, the successive values of A_i then being 1, 7, 19, 37, 61, 61, 37, 19, 7, and 1. The pattern of these numbers reflects the greater impor-

Table S6.5/1 100-second creep modulus at low strain; unplasticized PVC at 20°C

Deformation type	End condition	Modulus (GPa)
Cantilever	Screw-linked metal plates; low torque	2.78
	Screw-linked metal plates; high torque	2.90
	Cemented into metal block	3.06
Centre-loaded beam	Knife-edge supports	3.16

tance of the central region of an end-supported centre-loaded beam; the corresponding region for the cantilever is that nearest the clamping point, which partly explains the sensitivity of that method to end-effects.

The beam equations are strictly valid only if the properties of the material are constant through the thickness of the specimen, otherwise the calculated creep modulus is biased towards that of the skin layer. In contrast, tensile measurements give a simple mean of any through-thickness variation. Such skin-bias limits the immediate relevance of any flexure creep data to sections of similar thickness to that of the test specimens and of similar thermal history and has important implications for flexure testing of laminated samples, e.g. fibre composite structures. When thick specimens have to be tested, the span has to be increased in proportion if the accuracy is to be preserved and the apparatus may then be massive. A machine with adjustable supports giving a choice of span up to 600 mm has been developed by Mucci and Ogorkiewicz.[9] It has the refinements of a counterbalance for the loading shackles, a hydraulically damped loading mechanism and interchangeable attachments providing a choice between single knife-edge, two-knife-edge and single-point loading. The deflection of the beam can be resolved to within ±0.0025 mm which, in conjunction with the relatively large dimensions of the machine, confers a high overall accuracy. It is suitable for creep tests on beams or more complicated structures made from materials as diverse as thermoplastics and carbon fibre reinforced epoxy systems.

Shear

Various configurations are used for the measurement of shear modulus by ramp excitation but, with one exception, they have rarely been used in creep or relaxation tests. The common exception has been torsion but recently, also, Yoosefinejad and Hogg have developed and used a modified double Iosipescu test piece[10] and earlier Mallon and Benham described a shear creep apparatus,[11] which can be regarded as an infinitely long two-rail system. That method has not been used widely despite its obvious potential for high accuracy and high precision measurements. This is probably because of the exacting demands on the precision of machined test pieces and the initial requirement for relatively thick and uniform sheet stock, which limits the range of materials that can be tested by it.

A simple and inexpensive apparatus for the study of shear creep compliance via torsion was developed by Bonnin, Dunn and Turner.[12] By deliberate choice it uses the same test piece as that generally used in tensile creep studies. Its rectangular cross-section is not ideal for torsional deformation but, provided all the experimental strains are small and adequate recovery periods are allowed, one test piece can be successively stressed in torsion, flexure and tension, from which inferences can be drawn about its anisotropy, see also Chapter 7. Such multiple use of the test piece eliminates interspecimen variability from the inferences though even with an accuracy of ±1% on the flexural and torsional measurements, which is not easily achieved, the calculation of a notional Poisson's ratio is only accurate to about ±7%.

Solid rectangular prisms are not ideal for torsion because the strain varies over the cross-section, which complicates the analysis if the material is non-linear elastic or viscoelastic and because under torsion non-circular plane sections distort out-of-plane, which introduces tensile stresses. When the strains are low the following equation should suffice:

$$G(t) = \frac{LM_t}{\beta bh^3 \theta(t)} \qquad\qquad [S6.5.3]$$

where $G(t)$ = shear creep modulus
      L = length
      M_t = applied torque
      b = width
      h = thickness
      θ = angle of twist (in radians)
      β = $f(b, h)$, a shape factor.

Mixed mode tests

There have been various studies of creep under combined stresses, all seeking to clarify the underlying theory and some incorporating the additional complexities of non-linear superposition and complex stress histories. The test machines are inevitably elaborate and the experimentation is fraught with possibilities for error. The issues remain largely unresolved and work proceeds spasmodically and only in academic institutions.

References

1 C. M. R. Dunn. Private communication.
2 W. H. Mills and S. Turner. 'Tensile creep testing of plastics'. Paper No. 23, I. Mech. E. Symposium on Developments in Materials Testing Machine Design, Manchester, England, 7–10 September 1965.
3 M. W. Darlington and D. W. Saunders. 'An apparatus for the measurement of tensile creep and contraction ratios in small non-rigid specimens'. *Journal of Physics E.* 3, 511, 1970.
4 E. D. Jones, G. P. Koo and J. L. O'Toole. 'A Method for measuring compressive creep of thermoplastic materials.' *Material Research and Standards* 6, 241, 1966.
5 D. A. Thomas. 'Uniaxial compressive creep studies.' *Plastics and Polymers* 37, 485, 1969.

6 W. N. Findley and J. Gjelsvik. 'A versatile biaxial testing machine for investigations of plasticity, creep or relaxation of materials under variable loading paths.' *Proceedings ASTM*. 62, 1103, 1962.

7 P. J. Mallon and P. P. Benham. 'Effects of size and surface on the creep of polymers.' *Nature, Physical Science* 230, 2, 45, 1971.

8 C. M. R. Dunn and S. Turner. 'Modulus and creep in flexure of plastics: apparatus and Technique for Accurate Measurements.' *Plastics and Polymers* 40, 27, 1972.

9 P. E. R. Mucci and R. M. Orgorkiewicz. 'Machine for the testing of plastics under bending loads.' *Journal of Strain Analysis* 9, 3, 141, 1974

10 A. Yoosefinejad and P. J. Hogg. *A Long-term Shear Test for Orthotropic Composites*. Composites Testing and Standardisation ECCM-CTS-2, Ed P. J. Hogg *et al.* Woodhead Publishers, Abington, 1994, pp. 397–408.

11 P. J. Mallon and P. P. Benham. 'The development and results of a shear creep test for thermoplastics.' *Plastics and Polymers* 40, 22, 1972.

12 M. J. Bonnin, C. M. R. Dunn and S. Turner. 'A comparison of torsional and flexural deformations in plastics.' *Plastics and Polymers* 37, 517, 1969.

S6.6 Time–temperature superposition and stress relaxation master curves

Creep and stress relaxation studies usually require experiments of long duration, but in the case of relaxation most of the studies that have been reported have invoked time–temperature superposition to give ostensibly long-term data from short-term experiments. The method is based on the simplifying assumption that a change of temperature merely changes the molecular mobility on which creep and stress relaxation depend in such a way that each relaxation time is changed by the same factor and therefore a test at a higher temperature should identify a part of the relaxation spectrum that would be manifest only at longer times if the temperature were lower. A horizontal shift of the partial creep or relaxation curves, possibly with some second-order vertical shift, should allow the curves to be superposed into a 'master curve' extending over many decades of log time. For complex modulus the process is a frequency – temperature superposition, see Supplement S5.9.

There are two flaws in the argument. A change of temperature often entails a change in the molecular state of order in a plastic, with the complication that the equilibrium state of order is not attained instantaneously. Moreover, the rate of equilibration is of a similar order to the rate of change of modulus since the same segmental motions govern both changes and therefore state of order is likely to change during the test so that the test pieces used at the different temperatures cannot be regarded as nominally identical. The second flaw in the argument is that, at any one temperature, more than one class of molecular transition, associated with different activation energies, may be contributing to the observed creep or relaxation and, furthermore, the relative contributions will depend on the temperature locations of the transitions with respect to the test temperature.

There are regions of the time–temperature–modulus relationship where time–temperature superposition is an acceptable procedure, at least for stress relaxation, in particular where the glass–rubber transition is dominating the properties. In that region, relaxation modulus decays relatively rapidly with respect to log time and the lateral shifts needed to superpose the constituent curves are small. Furthermore, a measurement error corresponds usually to a vertical shift of the curve which, because of the slope, does not create a dominatingly large error in the horizontal shift needed to achieve an acceptable superposition overlap with neighbouring components of the master curve. Even so, the transition region is also the region where the state of molecular order is particularly sensitive to temperature and storage history. A check can easily be built into the experimental programme; a test at one temperature should be continued for 10 or preferably 100 times the duration of the other partial curves – the extended partial curve should coincide with the constructed master curve if the superposition is valid.

Time–temperature superposition seems to have been applied much more vigorously and successfully to stress–relaxation data than to creep data. One possible reason for this is that the objective of creep tests is mainly the prediction of long-term load-bearing capability, usually at temperatures well away from the glass–rubber transition temperature, where creep compliance is not likely to change rapidly with respect to log time. In those cases superposition requires large horizontal shifts and small measurement errors then correspond to large errors in the shift factor so that the whole superposition process is suspect at the manipulation level. Another reason for quantitative inadequacy of time–temperature superposition of creep data, as distinct from the unquantitative use mentioned in Section 6.6, is that viscoelastic non-linearity is equivalent to a distortion of the relaxation and retardation spectra, i.e. it contravenes the conditions for satisfactory time–temperature superposition; that can be accommodated within a series of relaxation experiments by the simple expedient of the strain being held constant throughout, which ensures that the degree of non-linearity remains approximately constant, whereas no such solution is possible during creep experiments. Even so, Larsen and Miller[1] had applied the technique to creep and creep rupture in metals, and Goldfein[2] subsequently did the same for plastics, equating the effect of time and temperature through a parameter K given by:

$$K = T(20 + \log t)$$

where T = absolute temperature in degrees Fahrenheit
t = time in hours

The technique was claimed to offer adequate accuracy for relatively rigid materials, such as glass fibre reinforced resins, in both tension and flexure, though the reported discrepancies between calculation and experiment were of the order of 20%. It should be noted that such materials are only usable at low strains, where they are approximately linear in viscoelasticity

and where $\dfrac{d\varepsilon}{d\log t}$ is small.

References

1 F. R. Larsen and J. Miller. 'A time–temperature relationship for rupture and creep stresses.' *Transactions of the American Society of Mechanical Engineers* 74, 765, 1952.

2 S. Goldfein. 'Prediction techniques for mechanical and chemical behaviour.' Ch. III. *Testing of Polymers* V.4. ed. W. E. Brown. Interscience, 1969.

S6.9 Recovery after creep

The recovery experiment is basically simple; after a period of creep, the applied force is removed smoothly and rapidly. The residual strain decreases with the passage of time, along a course which is essentially a creep curve in reverse. It is an important adjunct to creep studies, because recovery during unstressed periods in the service lifetime of a load-bearing item reduces the overall creep strain relative to that reached under uninterrupted loading. Additionally, in certain cases it distinguishes between reversible and irreversible creep strains. Even so, recovery has been studied less intensively than creep and the patterns of behaviour have not been classified in detail, largely because of the effort and cost entailed in a thorough mapping of the behaviour.

The data can be plotted in several different ways and the choice depends on the ultimate use to which they are to be put. The time origin may be taken as the moment at which the preceding creep period started or as the moment at which the recovery period itself started. The alternative visual displays are very different when the elapsed time is plotted on the usual logarithmic scale.

Discrepancies arise between the observed residual strain and that predicted by linear superposition for a combination of reasons, mainly viscoelastic non-linearity and progressive changes to the state of molecular order in the test piece during the course of the test or service lifetime. The former causes recovery to be greater than that predicted by linear superposition theory during the early stages of the recovery phase and less than that predicted during the later stages; the latter effectively changes the ground state of the test piece which cannot then revert to its original dimensions. Apart from those distorting influences, recovery experiments are prone to errors due to the restoring forces diminishing as recovery progresses to the point where they cannot overcome the small but inevitable frictional resistances in the mechanical system. Thus, what is observed may be only an approximation to the true recovery behaviour and even if there are no crude experimental errors such as the intrusion of untoward frictional resistance substantial levels of unrecoverable strain may signify a changed ground state rather than plastic deformation.

The quantities used as coordinates in Fig. 6.11/1 (main text) were chosen for pragmatic reasons in the light of the early results and the absence of a proven supporting theory. They are:

$$\text{Fractional recovery (F.R.)} = 1 - \frac{\text{residual strain}}{\text{maximum creep strain}}$$

$$\text{Reduced time}(t_R) = \frac{\text{recovery time}}{\text{preceding creep time}}$$

They have no recognized role within the various mathematical formulations that have been proposed but they enable the recovery function to be represented by a family of simple curves which in some limiting cases condense into a master curve. The trend is well illustrated in Fig. 6.11/1; the deviation from a common master curve becomes more pronounced as either the final creep strain or the creep duration increases. The recovery of thermoplastics after creep at elevated temperatures (but well below the glass–rubber transition) shows reduced tendency or no tendency to fall onto a common curve, see, for example, Fig. S6.9/1. It falls well below the recovery expected of thermoplastics at room temperature, mainly reflecting changes induced by elevated test temperatures in the state of molecular order and hence, as mentioned above, changes in the ground state. The most recent work on modelling recovery behaviour starts with the assumption that each change in the stress applied to a test piece initiates changes in the state of molec-

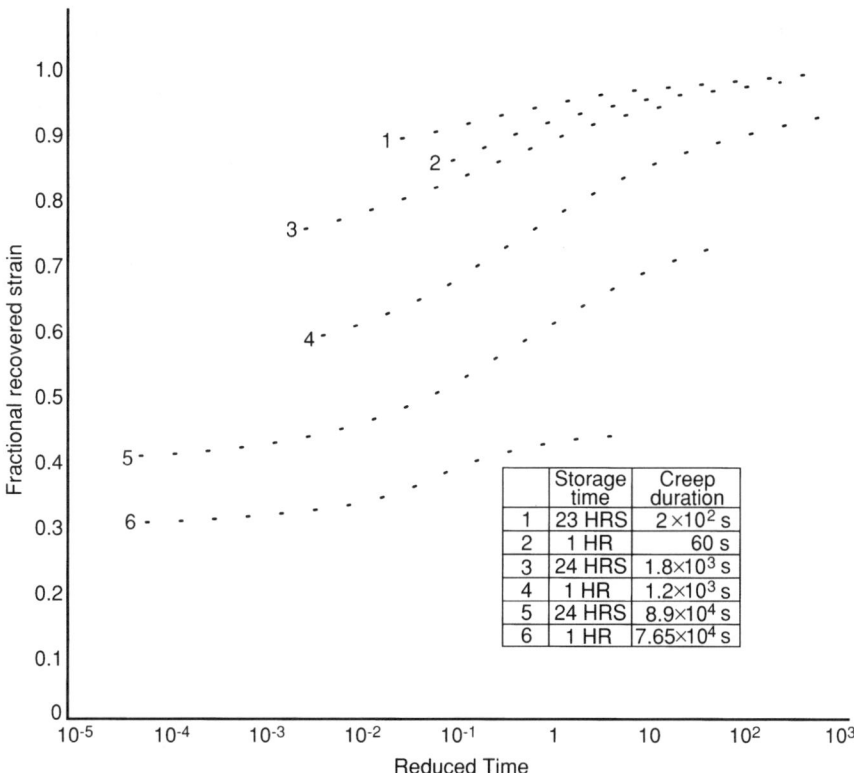

	Storage time	Creep duration
1	23 HRS	2×10^2 s
2	1 HR	60 s
3	24 HRS	1.8×10^3 s
4	1 HR	1.2×10^3 s
5	24 HRS	8.9×10^4 s
6	1 HR	7.65×10^4 s

S6.9/1. Recovery after creep. PVC at 60 °C. Strain recovery is partly suppressed if ageing occurs during the creep and recovery periods. Preconditioning at the test temperature reduces the effect.

ular order. The predictive value of that concept has not yet been quantified. Zapas and Crissman[1] had earlier reported success with what they termed plasto-viscoelasticity in the possibly special case of ultra-high molecular weight polyethylene. They had taken note of much earlier reports that creep and recovery data for natural polymeric materials (W. Weber 1835) and synthetic polymeric materials (H. Leaderman 1943) could be reproduced 'only when the specimens had been subjected to prior mechanical conditioning', all of which may be equivalent in ultimate effect to the latest ideas.

Apart from the unproven ability of current theories to account quantitatively for recovery behaviour, there are behavioural anomalies additional to those mentioned above, particularly in the recovery characteristics of glassy amorphous polymers. For instance, full recovery may be observed in specimens that have crazed during creep, even though crazes are discontinuities bridged by drawn material that cannot be expected to retract completely; however, the crazes tend to disappear as the strain decreases after the stress is removed. The early stages of stress whitening are similarly reversible but, on the other hand, the pronounced stress whitening that lies within Lüders bands originating at edge defects or other points of stress concentration do not disappear, and are associated with a significant level of unrecoverable strain.

Reference

1 L. J. Zapas and J. M. Crissman. 'Creep and recovery behaviour of ultra-high molecular weight polyethylene in the region of small uniaxial deformations.' *Polymer* 25, 57, 1984.

S6.10 The isochronous stress–strain procedure

The primary evaluation of the creep characteristics of a non-linear vis-coelastic material entails creep tests at several stress levels. However, in many cases the testing burden can be reduced by the use of a special experi-mental routine in which the test piece is subjected to a sequence of creep and recovery periods with the applied stress greater in each successive creep period. The duration of each creep period is short (commonly 100 seconds and sometimes 60 seconds) and the unstressed recovery period is suitably longer but the same in each cycle (four times the creep duration is usual but a greater factor may be needed if the recovery rate is particularly slow). The stress applied during each creep period plotted against the strain reached by the end of it is the isochronous stress–strain relationship at that time; it is a constant-time section across a family of creep curves and con-stitutes a simple index of non-linearity. It is also the spacing rule at that spe-cific time for the family of creep curves, in which role it reduces the number of experimental creep curves that are required to characterize the time dependence. Additionally, if the strain response during the course of each creep period is recorded, the data constitute a map of the creep behaviour up to the time of the isochronous section.

The experimentation procedure is simple in principle and execution. However, if high accuracy is sought, then the simplicity evolves into an elab-orate formality[1] to ensure that the creep during successive periods under stress is not affected by a number of factors. These include the stress history, the accumulating unrecovered strain and the extraneous friction in the machine (which would cause an apparent deviation from linearity at low levels of applied stress).

The applied stress plotted against the strain immediately prior to the removal of the stress is a straight line for a linear viscoelastic material and deviation of the data towards disproportionately greater strains is the usual manifestation of non-linearity. Another linearity criterion is a straight line of unit slope if log stress is plotted vs. log strain. The latter display mode is particularly useful because it has transpired that the shape of the curves is unique for each class of plastic and is also largely independent of thermal history and anisotropy. The effect of those variables is a displacement of the curve along the log stress axis. Consequently, a partial picture of the influence of those variables on the creep behaviour can be deduced from a few isochronous tests. Thus, important economies in testing costs can be achieved.

There is one exception to the phenomenon of shape constancy, namely the effect of an incorporated filler. The enhancement factor, i.e. the ratio of the creep moduli of the filled and the unfilled samples, usually decreases as the strain increases. The curves tend to converge with increasing strain, and in extreme cases the curve for the filled sample may even fall below that for the base polymer, but the same effect can be seen more expeditiously, though less precisely, on a conventional stress–strain or force–deformation curve.

Isochronous curves plotted as log stress vs. log strain generally lie on straight lines of slope slightly less than unity (generally about 0.95) up to a strain of about 0.005, and above that on curves of steadily decreasing slope. Those effects seem to be manifest with all plastics other than the most rigid ones and apparently with all models of high precision creep machines. Therefore, they seem to be authentic evidence of general non-linearity, though it must be borne in mind that friction in the loading mechanism could cause the same effect. Very similar results have been obtained in the less numerous experiments based on other deformation modes; for example, uniaxial compression data show the same type of non-linearity, so presumably there is a small linear region at the transition from tension to compression.

An isochronous procedure should be followed at the start of a creep evaluation programme to enable the investigator to make a judicious choice of stress levels for the creep tests. Furthermore, since there is inevitably some interspecimen variability even in samples of the highest quality there is need for a reference base to be established, for which purpose the isochronous procedure is ideal on account of its brevity and its in-built checks. Thus, the mean curve of several should be derived and the individual creep curves should be adjusted as necessary to coincide with it. Experience has shown that with well-regulated samples and specimens, disparities between the creep and the isochronous data should be no greater than 1% of the mean isochronous value. Larger disparities are usually indicative of apparatus malfunction, faulty test pieces or irregular samples, e.g. samples with poorly dispersed filler.

Reference

1 D. A. Thomas and S. Turner. 'Experimental technique in uniaxial tensile creep testing.' p73, *Testing of Polymers* ed. W. E. Brown. Interscience, 1969.

S6.14 Creep databases and testing strategy

Databases on the properties of plastics materials vary widely in contents, scope and purpose. The quality of the data also varies, partly with the purpose of the database and partly as an inevitable consequence of changing consensus as to what data are required, how they should be presented and how the communication pathways should operate. In the particular case of creep data for plastics the main purpose of the data generation is the provision of downstream support for the design of load-bearing end-products. Data generators have almost exclusively been materials producers and naturally it is they who have had the greatest influence on the contents, formats etc. of the current creep databases. A typical creep properties database currently consists of families of creep curves at several stresses measured at one or more of the internationally accepted standard temperatures or possibly at a temperature deemed to be especially relevant in the light of the pattern of usage of the material. Those data relate to a specific state of molecular order in the material and may be supplemented by limited information about the effects of state of order on the creep resistance. Recovery data are not yet widely included in the databases though they can have an important bearing on fitness-for-purpose of end-products.

It is clear that the establishment of a creep properties database, even within the current limited scope, is an expensive operation. If the generated data have to highlight ageing effects, possible anisotropy, sensitivity to processing conditions, etc. the cost burden may be unbearable. However, various procedures, especially the use of an isochronous stress-strain curve as a spacing rule and the identification of material-specific shapes of creep curves, can reduce that burden, as was discussed earlier in Chapter 6. Also, when a material is almost linear viscoelastic and used mainly in its glassy state a single creep curve may suffice in place of the family of curves that is otherwise necessary. Figure S6.14/1 shows an example where one curve at each of three temperatures probably suffices.

Apart from such possible economies in testing effort, there has also been some change of emphasis from data generation and processing to matters of presentation format, which may lead ultimately to a simplified data content and consequential reduction in the enabling experimentation. In the past, a plastics producer would measure the creep behaviour, publish the data as creep curves and expect a designer to extract the information that was needed to calculate adequate load-bearing cross-sections. Supplementary information enabling allowance to be made for differences in material state and other influential factors would be provided, normally via a technical service function. However, that function can be largely dispensed with if the database is expressed in terms familiar to a design engi-

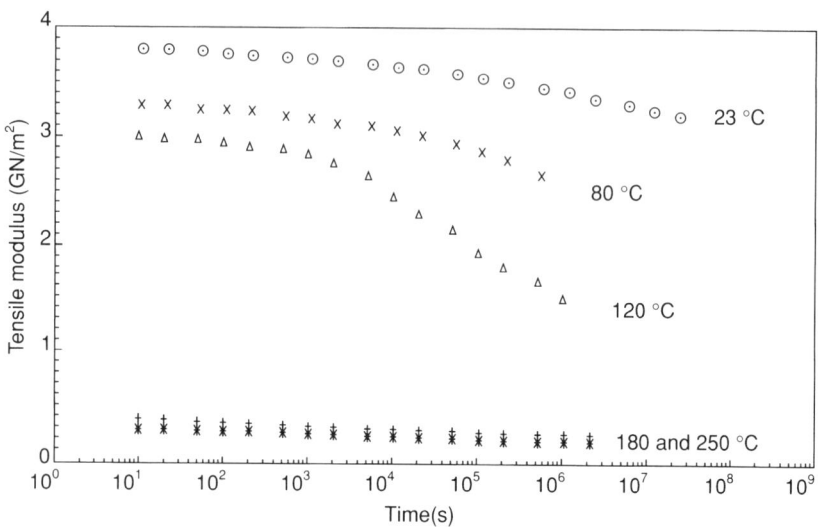

S6.14/1. Isometric tensile creep modulus of polyether ether ketone at various temperatures. The strain was 0.005. This display is the minimum for a creep database. Depending on circumstances, it may just suffice for an amorphous thermoplastic in its glassy state.

neer rather than those familiar to a polymer scientist. The following procedure is likely to be helpful in that respect:

1   Tensile creep data as a function of elapsed time and applied stress are interpolated by use of an isochronous stress–strain curve and extrapolated when conditions are favourable.

2   The processed data are converted into creep modulus and tabulated at suitable values of elapsed time. This modulus relates to either the applied stress (in which case the modulus is the reciprocal of the creep compliance) or the strain (in which case the modulus is derived from an isometric cross-plot as discussed elsewhere).

3   Additional data are generated at the highest temperature at which load-bearing service is anticipated and the manipulation procedures repeated. The effect of intermediate temperatures is estimated from a modulus vs. temperature curve obtained by sinusoidal excitation of a test piece taken from the same sample of material and using the same deformation mode.

The enabling science and the data adjustment techniques for those operations have been established and the requisite raw data for some plastics already exist in the archives. However, the various operations by which basic data may be manipulated and 'expanded' are likely to introduce errors or approximations that reduce the accuracy of the outcome. Fortunately, that degradation process does not automatically diminish the usefulness of the data, but obviously the testing strategy should be attuned to the possibility and must be adjusted as necessary to match the reliability of the final data to what is required for a satisfactory load-bearing design.

The array of creep modulus data can be rendered possibly more user-friendly by expressing the values as decimal fractions of a reference value of modulus. Supplementary data on the effects of anisotropy, thermal history, composition, etc. can similarly be reduced to decimal fractions which may, of course, be greater than unity. Reference 1 gives examples of what simplifications and condensations can be achieved in data presentation formats.

The quantity that should be used as the reference value is a matter for deliberation and discussion, but a secant modulus from a conventional tensile test at the international standard temperature and extension rate and at a strain of, say, 0.005 would probably be acceptable because it is so widely available. Alternatively, for data such as Fig. S6.14/1 the reference datum could be the 10 second modulus at 23 °C, which is essentially what the eye–brain combination does qualitatively. Depending on how the reference modulus is defined, the decimal fractions may be slightly less accurate than the modulus values that they replace but the relative values will be unaffected.

The subject of database formats and efficient communication pathways between data generators and data utilizers has been discussed at various forums and some local initiatives have created draft databases along similar lines to that mentioned above. However, development beyond such tentative steps cannot be rapidly consolidated by international standardization because that has to reflect a compromise between the often conflicting interests of the participating bodies. Compromise may be elusive in this particular case because the decimal fractions highlight what are sometimes severe reductions in the effective modulus, and there seems to be some hesitancy on the part of primary producers to collaborate with their competitors on such collective exposure.

There is another possible way forward which may be more attractive to the primary producers in that the harsh truth of modulus derating, which is quite inescapable, would be expressed less blatantly. That way forward requires the development of satisfactory, all-embracing creep equations. It is favoured by the finite element analysis community and offers the great potential advantage of data condensation. However, in the light of the very limited success of earlier creep equations, the complexity of the newer versions and the current absence of validation of those that have been proposed recently, this route does not offer a solution to the data condensation problem in the near future. Nor would creep equations offer substantial economies in testing except in the unlikely event of the discovery of a universal creep equation.

That evolving situation presents data generators with certain dilemmas over their future choice of testing strategy because expensive creep tests cannot be undertaken lightly when some of the results may subsequently be rendered redundant by future changes in the preferred format for creep databases. Thus, for instance, the development of a creep equation requires apparatus of the highest attainable accuracy, as well as tests on several samples with different processing histories and validation testing over the whole range of anticipated usage. But, if the final outcome is to be a simple array of decimal fractions, then a better testing strategy might be the use of one accurate machine to generate isochronous stress–strain data and many cheaper, less accurate machines to give average response curves for both standard test pieces and critical basic shapes. However, the data should always be scrutinized by the data generators for indicators of both machine malfunction and shortcomings of the material as a long-term load-bearer. This should be irrespective of whether the data are destined to be presented ultimately as a creep equation or as an array of modulus data (or decimal fractions thereof).

Reference

1 D. R. Moore and S. Turner. 'A modern strategy for the generation of creep data for plastics.' *Plastics and Rubber Composites: Processing and Applications* 21, 1, 33, 1994.

Modulus and stiffness anisotropy

7.1

Irrespective of the class of excitation, many modulus data for plastics reflect anisotropy and the relationships between force, deformation and modulus are correspondingly more complicated than those for isotropic elastic materials, see Supplement S7.1. Two independent moduli suffice to characterize the stiffness of an isotropic material, but up to 21 stiffness coefficients may be required for an anisotropic one; the coefficients are viscoelastic functions if the material is polymeric in nature.

In continuous and discontinuous fibre composites, anisotropy arises through the alignment and the spatial distribution of the fibres. In unreinforced plastics, it is a consequence of molecular alignments brought about as the molten polymer is processed into an end-product (the flow geometry effects), see Chapter 1, Section 1.5, Supplement S1.5 and later in this chapter.

7.2

In the composites industry there was always a strong awareness of the special problems attendant upon, and the special opportunities offered by, the anisotropy present in many fabricated structures. The associated complexities of evaluation, interpretation and design have been accommodated largely by empirical methods though there is also a sound theoretical background that had its origin in earlier technologies, e.g. fabricated wooden structures. In the thermoplastic industry, the special properties conferred by molecular orientations were readily exploited in films, tapes, etc. but the associated knowledge, particularly the awareness that an oriented system may have inferior properties in a transverse direction, were assimilated only belatedly by the other sectors of the industry. Thus, some of the properties data for thermoplastics published prior to 1975 are questionable in that adventitiously enhanced property values were published without recognition of and reference to the associated impairment of other properties. Furthermore, some of the test procedures of that time are still widely used and are potentially unreliable if test pieces are wrongly assumed to be isotropic.

7.3

The degree of anisotropy varies with the material and the processing route. For example, the ratio of maximum tensile modulus to minimum tensile modulus (anisotropy ratio) lies in the range 10–50 for a unidirectional continuous fibre composite. For other material types, this value will depend on volume fraction and perfection of alignment of the fibres. It will have a value of up to 10 in thermoplastic fibres and films, up to 3 in thermoplastics reinforced by short discontinuous glass fibres and up to about 1.3 in unreinforced thermoplastics processed without deliberate intent to orient the molecules.

7.4

In a thermoplastic moulded item the anisotropy ratio and the orientation axes vary from point to point and sometimes at first sight anomalously, with the axes of the anisotropy at variance with the nominal main flow axis, see Supplement 7.4, for instance. They vary also through the thickness, to such a degree that separate measurements of modulus by tension and by flexure give different maps of the anisotropy. The pattern of variation is governed by the interaction between the pressure source, the local impedances to the flow of the molten and solidifying plastic and the spatial and temporal variations in the rheological and thermal properties during the melt processing stage.

The variations generate uncertainty as to what value of modulus should be used in a particular design calculation. This is because property data relating to specific test pieces may be inappropriate unless there is some close correspondence between the flow geometries of the test pieces and the prospective end-product. The uncertainty will lessen if computational methods similar to those used for continuous fibre reinforced composites can be validated for the less regular anisotropy patterns in melt-processed items.

7.5

For a continuous fibre reinforced composite the Young's situation is less confused; a parallel array of fibres gives a composite modulus, E_c, by the simple parallel model:

$$E_c = \phi E_f + (1 - \phi)E_m \qquad [7.5.1]$$

where subscripts c, f and m refer, respectively, to composite, fibre and matrix and ϕ is the fibre volume fraction.

Equation 7.5.1 is the Voigt equation, based on the assumption of uniform strain throughout the composite body. An assumption of uniform stress leads to the Reuss (or Kelvin) equation:

$$\frac{1}{E_c} = \frac{(1-\phi)}{E_m} + \frac{\phi}{E_f}$$

[7.5.2]

It has been shown rigorously[1] that $E_{c(\text{Voigt})} > E_c > E_{c(\text{Reuss})}$.

If the fibres are not perfectly aligned, one must introduce an inefficiency factor, η_a; if they are discontinuous another factor, η_d, must be used and if the interfacial bond is deficient a third factor, η_b, is needed. At the very simplest, one then has a relationship:

$$E_{\text{comp}} = \eta_a \eta_b \eta_d \phi E_f + (1-\phi)E_m$$

[7.5.3]

where η_a, η_b and η_d are difficult to isolate and measure and where two out of the three vary from point to point.

For random long fibres, the product of the three terms generally lies between 0.15 and 0.50 as the ratio of fibre modulus to matrix modulus varies between 100 and 10 and as ϕ varies from 0.1 to 0.6. For continuous non-parallel fibres, the product $\eta_a \eta_d$ has the value 1/6 for 3D random fibres, 1/3 for plane-random fibres and 1/2 for orthogonal fibres.

7.6

The reasoning embodied in Eq. 7.5.3 prevails for thermoplastics containing short discontinuous reinforcing fibres and for unreinforced plastics if local agglomerates of aligned molecules are regarded as reinforcing elements. A test piece or an end-product may be envisaged as consisting of lamellae within each of which are local, variously aligned, domains of parallel fibres or molecules. In principle, the through-thickness variations can be allowed for by an adaptation of the laminate analysis used for continuous fibre-reinforced composites.[2] However, that idealization is largely frustrated by a general dearth of information about the patterns of fibre and molecule alignments and by the occurrence of out-of-plane alignments which are common because the melt flow regime is seldom simply laminar.

In general, short fibre-reinforced thermoplastics exemplify the point of principle discussed earlier in the monograph, namely that the test pieces from which properties are derived are themselves end-products with their own flow geometries and concomitant point-to-point variability. Thus, any test yields a property of a structure rather than a property of a material and any anisotropy is a characteristic of that structure, not of the material.

7.7

One consequence of those lines of reasoning and of the high demand for suitable data for the pseudolaminate analysis was the development of an unorthodox test configuration for the determination of notional stiffness coefficients, Supplement S7.7. The modulus, stiffness coefficient or creep resistance is measured by flexure, but of either a square plate or a disc rather than of a narrow strip or beam. In the flexure tests the loading is usually three-line (analogous to the three-point loading of a beam); it may also be central or uniformly distributed but the former yields imprecise and variable data and the latter entails the use of a cumbersome test rig.

The line-loaded disc test is described in Supplement S7.7; it is particularly useful for the rapid assessment of in-plane anisotropy because rotation of the disc in relation to the loading system leaves the test geometry constant, apart from any small variations in disc thickness which can be readily allowed for.

7.8

The influence of processing variables on the stiffness of a thermoplastics product is indeed complex. The interactions between cavity geometry, gate geometry, melt viscosity and melt elasticity are so complex that a sustained and systematic investigation using test pieces from several sources is necessary if the critical facts are to emerge. A prime requirement of the test methods is then that they should be speedy, preferably with no sacrifice of accuracy. The plate tests are advantageous in that the test pieces can be directly moulded or easily cut from larger items and do not have to be machined to demandingly fine tolerances. Furthermore, surface features such as grooves and ridges, various gating systems and other features can be incorporated so that test results may reflect the likely stiffness of some items in service.

7.9

In general, the data produced by such methods are likely to be less precise than those obtainable from conventional test pieces because the shapes of the former generally differ from what is ideal for the measurement of modulus. However, precision is of dubious merit if the derived value does not reflect practical realities or if it is only one of many possible values. Therefore, it can be argued that, on balance, the use of unconventional test pieces of simple shape is preferable to the use of conventional test pieces. Evaluation on any test piece shape contributes information to a population of realizable property values. The systematic use of several shapes constitutes a major change in evaluation strategy. The emphasis moves away from isolated property values, narrowly defined by a specific test configuration and often a single flow geometry or composite structure, towards an assembly of property values loosely defined by a set of test configurations and their associated flow geometries or structures. This multishape specimen approach has prevailed in the fibre composites industry almost throughout its history because it has always used deliberate placement of the fibres to achieve particular properties and attributes in the final structure. The film and fibre sectors of the thermoplastics industry have similarly used molecular orientation for the attainment of particular properties, but in the bulk moulding sector the variations tend to be inadvertent rather than deliberate, though admittedly less dramatic in scale.

7.10

In the case of thermoplastics, the possible surfeit of data arising from a multiple configuration strategy could be condensed/simplified along similar lines to the methods used for creep data (Supplement S6.14). This would simplify the process of data selection for particular downstream purposes. Thus, modulus anisotropy data could be limited in the future to upper bound values (probably obtained by tests on end-gated injection-moulded bars) and anisotropy ratios derived from measurements on discs as described in Supplement S7.7. Table 7.10/1, from reference 3, gives some anisotropy ratios measured on injection-moulded samples of several thermoplastics. Many other results indicate that those values reflect the levels of anisotropy that commonly arise but some variation is to be expected, and the ratios for tensile deformations are also likely to differ from those for flexural deformations.

7.11

Strength anisotropy poses greater difficulties than modulus anisotropy for both data generators and data utilizers because *meso* and micro aspects of morphology associated with molecular and fibre orientation act as a strength deficiency along particular axes, which can induce a local failure. In addition, strength may also be affected adversely by flow irregularities, structural irregularities and other imperfections to a much greater degree than modulus is. In general, therefore, adequate evaluation of strength requires more tests than comparable evaluation of modulus (or stiffness coefficients). Even so, a single-datum strength test may nevertheless provide visual evidence of system anisotropy, for example by fibrillation, delamination, subsidiary cracking, whereas a corresponding modulus test cannot usually do so. There are also prefailure phenomena that provide

Table 7.10/1 Typical anisotropy ratios at 23 °C

Material and moulding type	Deformation mode	Ratio
Polypropylene, edge-gated disc	Beam flexure	1.03
Polypropylene, centre-gated tray	Tension	1.15
Polypropylene containing 0.19 weight fraction of glass fibres, centre-gated tray	Tension	0.75[b]
Polypropylene containing 0.25 weight fraction of glass fibres, edge-gated disc	Beam flexure	1.17
Polypropylene containing 0.25 weight fraction of glass fibres, double-feed weir-gated plate	Plate flexure	2.0
Polypropylene containing 0.25 weight fraction of glass fibres, asymmetric single-feed weir-gated plate	Plate flexure	1.4
Low density polyethylene, coathanger-feed weir-gated plate	Beam flexure	1.41
	Disc flexure	1.31
	Disc flexure	1.22[a]
Nylon 66, edge-gated disc (dry)	Disc flexure	1.0
Nylon 66, single-feed weir-gated plate (dry)	Plate flexure	0.94
Nylon 66 containing 0.33 weight fraction glass fibres (dry), edge-gated disc	Disc flexure	1.6
Nylon 66 containing 0.33 weight fraction glass fibres (dry), single-feed weir-gated plate	Plate flexure	1.8

[a] Different moulding machine.
[b] This result is an example of an odd, but explicable, consequence of flow geometry.

indicators of anisotropy (see Supplement S8.2) but are not clearly obvious at the low strains that are usually entailed in modulus tests.

References

1 Z. Hashin and S. Shrikman. 'A variational approach to the theory of the elastic behaviour of multiphase materials.' *Jounal of the Mechanics and Physics of Solids* 11, 127, 1963.
2 D. R. Moore, J. M. Smith and S. Turner. 'Engineering design properties for injection moulding compounds based on poly(ether ether ketone).' *Plastics, Rubber and Composites: Processing and Applications* 21, 1, 19, 1994.
3 D. R. Moore and S. Turner. 'A modern strategy for the generation of creep data for plastics.' *Plastics, Rubber and Composites: Processing and Applications* 21, 1, 33, 1994.

Supplements

S7.1 Force–deflection–modulus relationships in anisotropic systems

S7.1.1 Anisotropic elasticity

For anisotropic systems, the concept of modulus is replaced by that of stiffness coefficients. The basic assumption of linear elasticity theory is that the response to an excitation is a linear function of all the components of the excitation tensor. If the material is isotropic the relationships between the stress and the strain are relatively simple, but if the material is anisotropic the relationships are less straightforward and greater care has to be taken over nomenclature, definitions and the experimentation.

The general stress–strain relationship for an anisotropic elastic material is:

$$\sigma_{ij} = c_{ijkl}\, \varepsilon_{kl} \qquad\qquad [S7.1.1.1]$$

where σ_{ij} is the stress tensor
ε_{kl} is the strain tensor
c_{ijkl} are the stiffnesses (each suffix having possible integral values 1, 2, or 3)

Alternatively, $\varepsilon_{ij} = s_{ijkl}\sigma_{kl}$ [S7.1.1.2]

where s_{ijkl} are the compliances.

In the stress terms, the first subscript indicates the direction of the normal to the plane on which the forces act and the second subscript to the direction of the stress. The numbers 1, 2 and 3 are synonymous with x, y, z, respectively in the cartesian coordinate system. Therefore, $\sigma_{11}, \sigma_{22}, \sigma_{33}$ are normal stresses whilst $\sigma_{12}, \sigma_{23}, \sigma_{31}$ are shear stresses.

For reasons of symmetry in the stress and strain tensors, only 21 of the 36 stiffness or compliance coefficients are independent, and that number is reduced further if there are other symmetries in the material, i.e. if the material is not fully anisotropic, see Table S7.1.1/1. The most general case, with 21 independent elastic coefficients, is so complex as to be virtually unmanageable in both the analytical and the experimental aspects. However, that extreme situation rarely arises in practice because the fabrication processes impose some order, either from deliberate intent or inadvertently.

The case usually considered analytically is orthotropy, which is conferred by a multiangled symmetric array of fibres in a continuous fibre reinforced composite, e.g. a laminate comprising $[0, 90]_{2S}$. Orthotropy also applies to uniaxial drawing of fibres and films and by other directional processing of thermoplastics. Composite laminates, in which the lamellae are variously aligned with respect to one another but which have an unsymmetric stacking sequence, e.g. a laminate comprising $[0, +45, 90, -45]_8$, are analysed by a combination of orthotropy and laminate theory. In principle, that type of analysis should be feasible also for thermoplastic injection mouldings, which tend to be layered structures, but the molecular order is not as clearly developed in such systems as the fibre alignment in laminated composites.

Table S7.1.1/1 Classes of symmetry

Type of symmetry	Number of independent coefficients
None (triclinic)	21
One plane of symmetry (monoclinic)	13
Two planes of symmetry (orthotropic)	9
Transversely isotropic (one plane of symmetry)	5
Isotropic	2

The stress–strain relationship for an orthotropic system is:

$$
\begin{vmatrix} \sigma_{11} \\ \sigma_{22} \\ \sigma_{33} \\ \tau_{23} \\ \tau_{31} \\ \tau_{12} \end{vmatrix} =
\begin{vmatrix}
c_{11} & c_{12} & c_{13} & 0 & 0 & 0 \\
c_{21} & c_{22} & c_{23} & 0 & 0 & 0 \\
c_{31} & c_{32} & c_{33} & 0 & 0 & 0 \\
0 & 0 & 0 & c_{44} & 0 & 0 \\
0 & 0 & 0 & 0 & c_{55} & 0 \\
0 & 0 & 0 & 0 & 0 & c_{66}
\end{vmatrix}
\begin{vmatrix} \varepsilon_{11} \\ \varepsilon_{22} \\ \varepsilon_{33} \\ \gamma_{23} \\ \gamma_{31} \\ \gamma_{12} \end{vmatrix}
\qquad [\text{S7.1.1.3}]
$$

The strain–stress relationship is similar and is the more usual form, i.e.

$$
\begin{vmatrix} \varepsilon_{11} \\ \varepsilon_{22} \\ \varepsilon_{33} \\ \gamma_{23} \\ \gamma_{31} \\ \gamma_{12} \end{vmatrix} =
\begin{vmatrix}
s_{11} & s_{12} & s_{13} & 0 & 0 & 0 \\
s_{21} & s_{22} & s_{23} & 0 & 0 & 0 \\
s_{31} & s_{32} & s_{33} & 0 & 0 & 0 \\
0 & 0 & 0 & s_{44} & 0 & 0 \\
0 & 0 & 0 & 0 & s_{55} & 0 \\
0 & 0 & 0 & 0 & 0 & s_{66}
\end{vmatrix}
\begin{vmatrix} \sigma_{11} \\ \sigma_{22} \\ \sigma_{33} \\ \tau_{23} \\ \tau_{31} \\ \tau_{12} \end{vmatrix}
\qquad [\text{S7.1.1.4}]
$$

where the s_{ij} are the compliance coefficients. In Eqs. S7.1.1.3 and S7.1.1.4, ε_{11}, ε_{22} and ε_{33} are the principal strains and σ_{11}, σ_{22} and σ_{33} are the principal stresses. s_{qq} ($q = 4, 5$ or 6) relates a shear strain to a shear stress in the same plane.

The compliance matrix can be rewritten in terms of the more familiar engineering/physical constants E_{ii}, ν_{ij} and G_{ij}:

$$
\begin{vmatrix}
1/E_{11} & -\nu_{21}/E_{22} & -\nu_{31}/E_{33} & 0 & 0 & 0 \\
-\nu_{12}/E_{11} & 1/E_{22} & -\nu_{32}/E_{33} & 0 & 0 & 0 \\
-\nu_{13}/E_{11} & -\nu_{23}/E_{22} & 1/E_{33} & 0 & 0 & 0 \\
0 & 0 & 0 & 1/G_{32} & 0 & 0 \\
0 & 0 & 0 & 0 & 1/G_{31} & 0 \\
0 & 0 & 0 & 0 & 0 & 1/G_{12}
\end{vmatrix}
\qquad [\text{S7.1.1.5}]
$$

For a thin laminate, e.g. a single lamella, plane stress is often assumed, i.e.

$$\sigma_{33} = 0, \ \tau_{23} = 0, \ \tau_{31} = 0,$$

$\varepsilon_{33} = s_{33}\sigma_{33} + s_{32}\sigma_{22}$ and is therefore not an independent coefficient. Therefore, Eq. S7.1.1.4 reduces to:

$$
\begin{vmatrix} \varepsilon_{11} \\ \varepsilon_{22} \\ \gamma_{12} \end{vmatrix} =
\begin{vmatrix}
s_{11} & s_{12} & 0 \\
s_{21} & s_{22} & 0 \\
0 & 0 & s_{66}
\end{vmatrix}
\begin{vmatrix} \sigma_{11} \\ \sigma_{22} \\ \tau_{12} \end{vmatrix}
\qquad [\text{S7.1.1.6}]
$$

Alternatively,

$$
\begin{vmatrix} \sigma_{11} \\ \sigma_{22} \\ \tau_{12} \end{vmatrix} = \begin{vmatrix} c_{11} & c_{12} & 0 \\ c_{21} & c_{22} & 0 \\ 0 & 0 & c_{66} \end{vmatrix} \begin{vmatrix} \varepsilon_{11} \\ \varepsilon_{22} \\ \gamma_{12} \end{vmatrix}
$$

[S7.1.1.7]

where

$$
\begin{aligned}
c_{11} &= E_{11}/(1 - \nu_{12}\nu_{21}) \\
c_{12} &= \nu_{12}E_{22}/(1 - \nu_{12}\nu_{21}) = \nu_{21}E_{11}(1 - \nu_{12}\nu_{21}) \\
c_{22} &= E_{22}/(1 - \nu_{12}\nu_{21}) \\
c_{66} &= G_{12}
\end{aligned}
$$

An important feature of Eqs. S7.1.1.3, S7.1.1.4, S7.1.1.6 and S7.1.1.7 is that the normal and the shear components are uncoupled, i.e. normal stresses do not induce shear strains and shear stresses do not induce normal strains. That situation prevails only when the coordinate system for the stress field coincides with the symmetry axes. For a lamella whose material axes are oriented at an angle θ in the 1–2 plane the relationship in Eq. S7.1.1.6 is changed to:

$$
\begin{vmatrix} \varepsilon_x \\ \varepsilon_y \\ \gamma_{xy} \end{vmatrix} = \begin{vmatrix} \overline{s_{11}} & \overline{s_{12}} & \overline{s_{16}} \\ \overline{s_{21}} & \overline{s_{22}} & \overline{s_{26}} \\ \overline{s_{61}} & \overline{s_{26}} & \overline{s_{66}} \end{vmatrix} \begin{vmatrix} \sigma_x \\ \sigma_y \\ \tau_{xy} \end{vmatrix}
$$

[S7.1.1.8]

where $\overline{s_{11}} = \cos^4\theta + \cos^2\theta\sin^2\theta(2s_{12} + s_{66}) + \sin^4\theta s_{22}$

where $m = \cos\theta$
$n = \sin\theta$

The other transformed compliance coefficients have a similar form. The corresponding transformed stiffness coefficient is:

$$
\overline{c_{11}} = \cos^4\theta c_{11} + 2\cos^2\theta\sin^2\theta(c_{12} + 2c_{66}) + \sin^4\theta c_{22}
$$

The expression for $\overline{s_{11}}$ may be rearranged into:

$$
\frac{1}{E_\theta} = \frac{\cos^4\theta}{E_0} + \left(\frac{4}{E_{45}} - \frac{1}{E_{90}} - \frac{1}{E_0}\right)\sin^2\theta\cos^2\theta + \frac{\sin^4\theta}{E_{90}}
$$

[S7.1.1.9]

to give the in-plane variation of tensile modulus for this simple system of anisotropy. Raumann and Saunders[1] showed that Eq. S7.1.1.9 provided a good representation of the tensile modulus of strips cut at various angles from polyethylene sheets with forward draw ratios of up to 4.65:1. Other similar work followed (for instance, that of Ogorkiewicz and Weidmann[2]), and the equation has been widely used since then as an approximate representation of the in-plane anisotropy of injection moulded plates and other systems.

In cases such as that represented by Eq. S7.1.1.8, s_{qr} ($q = 1, 2$ or 3; $r = 4$, 5 or 6) relates an extensional strain to a shear stress in the same plane and *vice versa*. The normal and shear modes are then coupled, i.e. a tensile stress induces some shear and a shear stress induces some tensile strain. In the most general case, s_{qr} ($q, r = 4, 5$ or 6) relates a shear strain in one plane to a shear stress in a perpendicular plane.

The consequences of Eq. S7.1.1.7 and similar ones are important, namely:

If the principal axes of the stress field do not coincide with the symmetry axes of the test piece, extraneous forces and deformations will arise. If a laminate consists of lamellae with their symmetry axes variously aligned with respect to each other, then an in-plane stress field will give rise to stress discontinuities at lamellae interfaces and extraneous deformations. The severity of the effects will depend on the stacking sequence, the degree of asymmetry, the test modes, the clamping arrangements, etc.

For a viscoelastic material in an anisotropic state, the stiffness and compliance coefficients have to be seen as functions of time, temperature and strain, with corresponding implications for what is entailed in a comprehensive evaluation. Allowance has to be made for anisotropy, irrespective of the class of excitation. For continuous fibre reinforced composites, only those coefficients that relate to the behaviour of the matrix under stress are likely to vary with time under load, but that is sufficient for some composite structures to creep significantly under some stress fields and/or for extraneous deformations to arise. In contrast, all of the coefficients for a thermoplastic material creep and exhibit other viscoelastic characteristics, though to varying degrees. Fortunately, the full complexity of the viscoelastic functions and their interactions may need to be invoked only rarely, and usually it will suffice for data to be derived and handled in the simple time-dependent elastic form $c_{ij}(t, T, \varepsilon)$.

The point-to-point variation in the $c_{ij}(t, T, \varepsilon)$ that is common in thermoplastic mouldings limits applicability of laminate theory but, by the same token, it also reduces the mismatch between neighbouring notional laminae by the variation within each lamellar plane and the presence of out-of-plane molecular alignments.

It is possible for anisotropy to be regarded as the general case, with isotropy as a convenient approximation in some instances. However, that would be at variance with the route by which the mechanical testing of plastics developed. Nevertheless, the test procedures needed for anisotropic test pieces are essentially the same as those used for isotropic ones, except for certain restrictions on their shapes, e.g. the ratio of length to thickness or width. However, there is a marked disparity between the requirements for data as dictated by the formal rationale and what is actually provided.

Irrespective of what level of curtailment is imposed on a test programme, each datum should be qualified. This should include a statement specifying the test configuration, the anisotropy axes of the test piece and the imperfections inherent in the processing/fabrication stage. Even where anisotropy is deliberately imposed via the fabrication process, variations occur, especially if the imposition mechanism is itself unsymmetrical. For example, the axes of orientation symmetry in biaxially drawn flat film are constant only along one direction (which can be regarded as the main draw axis) and tend to vary systematically along the orthogonal direction, whereas they are essentially constant along the orthogonal direction if production is by the blown-film process.

References

1 G. Raumann and D. W. Saunders. 'The anisotropy of Young's Modulus in drawn polyethylene'. *Proceedings of the Royal Society* 77, 1028, 1961.
2 R. M. Ogorkiewicz and G. Weidmann. 'Anisotropy in polypropylene injection mouldings.' *Plastics and Polymers*. 40, 337, 1972.

S7.1.2 St Venant's Principle

Anisotropy influences the experimental procedures additionally via a secondary effect embodied as St Venant's Principle, which was enunciated originally for isotropic materials. In its original form, the Principle states that any differences in the stress states produced by different but statically equivalent load systems decrease with increasing distance from the loading points. They become insignificant at distances greater than the largest dimension of the area over which the loads are acting.

In an anisotropic specimen the uniform stress state is approached much more gradually. It has been shown that the decay length λ is of the order of:

$$\lambda = b\left(\frac{E_{11}}{G_{12}}\right)^{1/2} \qquad [S7.1.2.1]$$

where b is the maximum dimension of the cross-section.

For rectangular strips subjected to traction at the ends:

$$\lambda \approx \frac{b}{2\pi}(E_{11}/G_{12})^{1/2} \qquad [S7.1.2.2]$$

where λ is the distance over which a self-equilibrated stress applied at the ends decays to $\frac{1}{e}$ of its end-value.

The ratio E_{11}/G_{12} has a value lying between 40 and 50 for a unidirectional continuous carbon fibre reinforced composite with a fibre volume fraction of 0.6. The value would be about 3 for an isotropic specimen and if Eq. S7.1.2.1 is valid for the isotropic case, as it should be, the respective decay lengths are in the ratio of about 3.5:1. The practical consequences are that 'end-effects' and other disrupting factors can be more severe and/or more persistent in anisotropic systems than in isotropic ones. Special test piece profiles may have to be used to avoid untoward effects, such as premature failure in tensile tests or unusual failure modes. Thus, for example, reinforcing end-plates are virtually obligatory for the tensile testing of continuous fibre reinforced composites, with different stipulations on specimen size for each class of composite laminate, see Chapter 4.

Special provisions usually serve their intended purpose, but can simultaneously be troublesome by inadvertently introducing other deficiencies. For instance, the end-plates eliminate both the St Venant effect and any tendency for failure at the grips. However, they add to the cost of specimen preparation and also have to be precisely shaped and symmetrically aligned on the test piece, because otherwise they introduce axial misalignment and the associated measurement errors.

S7.1.3 Elementary formulae

It follows from the previous Supplement that even casual anisotropy may invalidate some of the assumptions on which the elementary working formulae are based. A simple example is the bending of a beam under three-point loading. The deflection is due to a combination of longitudinal strain and shear strain. For an isotropic system, the magnitudes of the relevant moduli are such that the contribution of shear deformation is negligible provided the specimen conforms to certain restrictions on dimensions. In some anisotropic situations, however, the ratio of Young's modulus to shear modulus is much larger than it is for an isotropic and homogeneous specimen and a more elaborate experimental procedure is then required. The simple formula for three-point bending:

$$\delta = \frac{PL^3}{4bh^3E} \qquad [S7.1.3.1]$$

has to be replaced by:

$$\delta = \frac{PL^3}{4bh^3E}\left[1 + \frac{3Eh^2}{2GL^2}\right] \qquad [S7.1.3.2]$$

where δ = deflection
    P = force
    L = span
    b = width

h = thickness
E = Young's modulus
G = shear modulus

The effect of a high E/G ratio can be mitigated by an appropriate choice of h/L. Alternatively, tests at various h/L enable both E and G to be derived, from a graph of δ/PL vs. L^2.

More rigorous solutions give different values for the constant factor in the shear term in Eq. S7.1.3.2. Wagner *et al.*[1] cited four references giving values 1, 1.2, 1.5 and $(12 + 11v)/10(1 + v)$, and their own experiments suggested an effective value of 1.18.

These principles are particularly important in the toughness measurements (see later chapters) of continuous fibre reinforced composites. For example, in mode II and mixed mode (i.e. a combination of modes I and II), delamination tests on unidirectional fibre reinforced composites, it is necessary to measure modulus in bending in order to progress the analysis of the fracture data.[2] The test pieces available for such modulus measurements may be relatively small, and an error in the toughness determination is then likely because of error in the modulus determination. Thus, in some recent work by the European Structural Integrity Society (ESIS) Polymers and Composites Technical Committee (TC4) on developing fracture test protocols, the following results (Table S7.1.3/1) were obtained for the determination of modulus of a unidirectional continuous glass fibre reinforced epoxy laminate, by application of Eq. S7.1.3.2.

Table S7.1.3/1 Modulus as a function of span for three-point bending data for continuous glass fibre reinforced epoxy resin

Span (mm)	Modulus (GPa)
50	8.8
70	12.1
90	13.7
Extrapolated to infinite span	18.9

It can be seen from these results that an accurate direct measurement of modulus in this three-point bending test is not possible unless the shear contribution to bending has been eliminated.

References

1 H. D. Wagner, S. Fisher, I. Roman and G. Marom. 'The effect of fibre content on the simultaneous determination of Young's and shear moduli of unidirectional composites.' *Composites* 12, 257,1981.

2 P. Davies, B. R. K. Blackman and A. J. Brunner. 'Towards standard fracture and fatigue test methods for composite materials'. ESIS Conference, Sheffield, 1998.

S7.1.4 The tensile modulus/shear modulus ratio

The polymer molecules are often oriented and extended during injection moulding or other processing procedures and other larger, ordered, structural entities may develop during the cooling phase. Those features confer anisotropy and, at the molecular level, inhomogeneities. The formal theory defines the elaborate system of tests that are necessary to quantify the mechanical behaviour of such samples accurately but a simple system of evaluation may suffice for many purposes and may be the only recourse when approximate indicators are required at short notice.

A simple anisotropy evaluation procedure was developed by Bonnin *et al.*[1] who studied the anisotropy of various plastics mouldings by sequentially measuring the 100-second tensile modulus, $E_c(100)$, by flexure and the 100-second shear modulus, $G_c(100)$, by torsion on single specimens (see Chapter 6, Supplement S6.5). In outline, the method considers the relationship between tensile modulus and shear modulus through the elastic relationship:

$$\frac{E_c(100)}{G_c(100)} = 2(1 + v) \qquad [S7.1.4\ 1]$$

The lateral contraction ratio, v, can be expected to have values in the range 0.3–0.4 for an isotropic material. Therefore, an isotropic material will have a moduli ratio of about 2.6 to 2.8. Consequently, values above this range will provide an indication and an approximate measure of the anisotropy.

The use of one specimen for the two measurements removes inter-specimen variability as a possible source of error and restrictions on the test procedure (see next paragraph) ensure that the derived moduli ratio will be sufficiently accurate to serve as an index of anisotropy.

The procedure was originally developed in the context of creep studies, specifically as a supplement to tensile creep. Under the particular test conditions that were imposed, a specimen could be tested first in flexure and then in torsion. In addition and following these tests, a tensile isochronous stress–strain evaluation and finally a creep test under sustained tensile stress can be carried out without any of the results being affected by the previous excitation(s). This is because the flexural and torsional measurements and the subsequent isochronous procedure were restricted to small strains and short times under stress.

At that time, there was a growing requirement for information on a second viscoelastic function for use in pseudoelastic design calculations. There was also some concern that anisotropy in test specimens might bias measured tensile creep data as well as invalidate the traditional pseudo-elastic design procedures. Some typical results, including very high ratios which justify the concern, are given in Table S7.1.4/1. In retrospect, values greater than 3 are to be expected in end-gated injection-moulded bars of plastics containing short glass fibres and lower, but still artificially high, values are likely to be manifest by bars moulded from unreinforced grades

Table S7.1.4/1 Shear and tensile creep moduli at 100 s, 20 °C and small strains

Material	Shear creep modulus[a] (GPa)	Tensile creep modulus[a] (GPa)	$\dfrac{E_c(100)}{G_c(100)}$	[b]Notional Poisson's ratio
Polymethyl methacrylate	1.17	3.12	2.66	0.33
	1.15	3.05	2.66	0.33
Polypropylene of density 910 kg/m^3	5.52	1.51	2.73	0.36
(from centre of 37 mm thick block)	5.62	1.56	2.77	0.39
Polypropylene of density 907 kg/m^3	5.26	1.59	3.02	(0.51)
(injection mouldings)	5.16	1.59	3.09	(0.54)
	5.41	1.64	3.02	(0.51)
Poly 4-methyl pentene-1 (injection moulding)	4.75	1.59	3.36	(0.68)
Polyethylene of density 955 kg/m^3 (injection moulding)	1.05	2.68	0.34	
Oxymethylene copolymer (injection mouldings – dry)	0.90	2.52	2.79	0.39
Oxymethylene homopolymer	1.15	3.24	2.82	0.41
Nylon 66	3.45	9.93	2.88	0.44
Nylon 66 + 0.33 weight fraction of glass fibre (injection moulding 65%rh)	8.84	6.26	7.08	(2.54)
PVC unplasticized				
From centre of 30 mm thick block	1.14	3.10	2.81	0.40
From 2.5 mm thick sheet	1.17	3.21	2.74	0.37
Oak (specimen axis along the grain)	8.48	9.07	10.7	(4.35)
Carbon fibre reinforced epoxy resin	3.88	176	45.3	(21.7)

[a] Creep modulus is defined as applied stress/consequential strain.
[b] Where the values of notional Poisson's ratio are obviously fictitious, they are in brackets.

of plastics such as polypropylene and poly(4-methyl pentene-1). Subsequently, alternative test methods and other ways of presenting data were developed, see Supplements S7.4 and S7.7.

Reference

1 M. J. Bonnin, C. M. R. Dunn and S. Turner. 'A comparison of torsional and flexural deformations in plastics.' *Plastics and Polymers* 37, 517, 1969.

S7.4 Anisotropy derating factors

It is a well-established practice for the tensile modulus (and the tensile strength) of thermoplastic compounds to be measured by tests on injection-moulded end-gated bars. It is nevertheless now widely recognized that such data vary with cavity shape, gating arrangement and processing conditions. Also, the data usually differ from those derived from specimens cut from other moulded shapes. A thin end-gated bar usually has a pre-dominantly axial flow geometry and therefore, depending on the class of polymer, the modulus value derived therefrom may be untypically high. It is clear that the traditional use of an end-gated bar in isolation can be seriously misleading about the modulus manifest in many service items. However, the method and the data have an entrenched status that is resistant to change, partly because of the pragmatic usefulness of the extensive archive data for end-gated bars and partly because alternative practices would entail higher costs. On the other hand, the disparities between data from end-gated bars and the properties of other moulded shapes may cause errors in design calculations leading to service malfunction.

A range of values of modulus can be expected in general mouldings made from some plastics and Table S7.4/1 gives examples, taken from reference 1, of the effect of flow geometry on the short-term tensile modulus of a glass fibre reinforced polypropylene. The results are expressed as derating factors relative to the short-term creep modulus of a standard 3 mm thick end-gated bar.[1] The factors for measurements in flexure are different but show similar variations. An alternative reference value could be the tangent modulus or the small strain secant modulus derived from an instrumented tensile test.

In principle it should be possible to predict the molecular order (and hence the mechanical properties) from a knowledge of the geometry of an item, the processing conditions and the rheological characteristics of the molten plastic and thereby calculate the derating factors. In practice, that is not feasible, partly because of the complexity of the physical situation and partly because currently the rheological properties are seldom evaluated in sufficient detail. A different approach requires fairly comprehensive data for a reference sample and additional, but relatively sparse, data for other subsidiary samples. The reference sample can be similar to, or even identical with, those traditionally used. The subsidiary samples should be so chosen as to relate loosely to classes of service item by featuring specific flow regimes and flow irregularities. The intention is to represent service items loosely rather than closely simulate them. The matter has not yet been pursued in depth, because of considerations about the interaction between single-point, multipoint and 'design' data. In addition, there are also con-

Table S7.4/1 Modulus anisotropy derating factors for various injection-mouldings; polypropylene reinforced by 0.25 weight fraction of short glass fibres

Cavity geometry and feed system	Stress direction relative to main flow axis	Thickness (mm)	Modulus derating factor
End-gated bar or strut	0°	3	1.00
		6	0.85
		12	0.45
Single feed weir-gated plaque	0°	3	0.90
	90°	3	0.60
	0°	6	0.85
	90°	6	0.50
Centre-gated radial flow disc	Radial	3	0.55
	Tangential	3	0.80
	Radial	6	0.50
	Tangential	6	0.70
Double end-gated bar[a]	0°	3	0.60
Double feed weir-gated plaque[b]	0°	3	1.00
	90°	3	0.50

[a] The mechanical properties at and near the opposing flow knit-line depend strongly on the flow length and the venting.
[b] The mechanical properties orthogonal to the merging-flow knit-line vary with distance from the gate.

cerns about what should be standardized and what should be a candidate for the role of reference data.

It can be argued that there is now a need for the most important classes of flow geometry to be identified and incorporated into novel test pieces that could be standardized and designated as 'critical basic shapes'; see Supplement S1.11. These should be intended to simulate end-products or parts of end-products without total loss of generality and without the physical integrity of the results being compromised. Tests that use certain simple 'critical basic shapes' are discussed in Supplement S7.7.

The plastics industry has used critical basic shapes for many years, particularly in impact test programmes, but has hitherto limited their use to *ad hoc* experimentation and avoided formal acknowledgement of their potential importance in a testing strategy. Stephenson's sources of derated modulus values[1] were similarly simple critical basic shapes in that they incorporated practical flow features.

The fibre composites industry has similar problems with the generation of service-pertinent data and employs a combination of special testing and computation. The latter has been particularly successful for continuous-fibre laminates which are relatively orderly systems compared with injection moulded thermoplastics items.

Reference

1 R. C. Stephenson. 'The acquisition and presentation of design information for thermoplastics composites – coupled glass-fibre reinforced polypropylene.' *Plastics and Rubber: Materials and Applications* 4, 45, 1979.

S7.7 Unorthodox test configurations

Flexure

The unorthodox tests in flexure differ from their conventional counter-parts mainly in the use of unusual test configurations and test pieces with molecular and fibre alignments similar to those commonly found in end-products and irregularities such as knit lines and manufacturing imperfections. The unusual test configurations are attained readily via attachments to standard test machines and plate-like test pieces with features such as grooves and flow-dividers offer numerous options on shapes pertinent to service items. Any of the excitation functions may be used, but the particular results cited below were derived from ramp excitations. However, subsidiary experiments have confirmed that, for instance, creep data obtained from plaques and discs are similar to those from conventional test pieces apart from the shape factors.

The use of a moulded plate rather than a beam cut from that plate simplifies the measurement of modulus in flexure by eliminating the specimen-machining stage and by providing in one measurement an average of the individual beam values across the plate. Figure S7.7/1 shows a simple plate flexure subassembly for use on a conventional testing machine and a set of typical results is given in Table S7.7/1; plate flexure stiffness (designated here by C in accordance with the nomenclature of anisotropic elasticity and not to be confused with $C_c(t)$, the creep compliance function) and anisotropy ratios for plates of a glass fibre reinforced polypropylene moulded in three different cavities shows how sensitive the data are to the geometry of the cavity and the gate.

Table S7.7/1 Effective plate stiffness and anisotropy ratio; different cavities; glass fibre reinforced polypropylene

Specimen	C (0°) (GPa)	C (90°) (GPa)	Anisotropy ratio
3.1 mm thick, flash-gated plaque, double-feed	6.3	3.1	2.0
3.1 mm thick, flash-gated plaque, asymmetrical single-feed	5.8	4.1	1.4
4.0 mm thick, flash-gated plaque, coat-hanger feed	5.5	3.3	1.6

$C(0°)$ is the plate stiffness along the main flow axis and $C(90°)$ is the stiffness in the orthogonal direction; it is equal to $E_{11}/(1 - v^2)$. In other nomenclatures these can be referred to as $C(L)$ and $C(T)$, respectively.

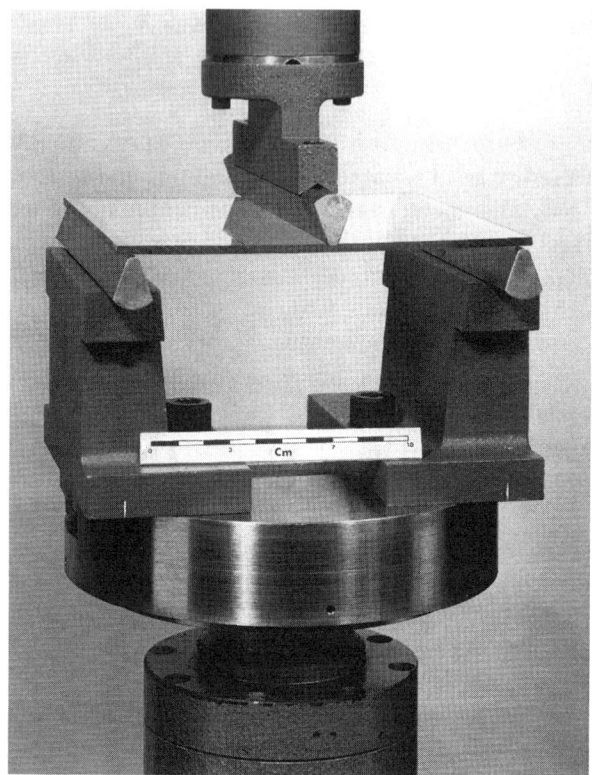

S7.7/1. Simple sub-assembly for the flexure of square plates. Provided that the maximum deflection in the first test is small, an accurate second test, with the plate flexed along the orthogonal direction, can be carried out. An anisotropy ratio can then be derived, see Table S7.7/1.

Corresponding values for a nylon 66 containing 0.33 weight fraction of short glass fibres were 9.8, 5.5 and 1.8, respectively, at 23 °C and 6.3, 2.5 and 2.5 at 80 °C. The increased anisotropy ratio at the higher temperature reflects an increased influence of the fibre orientation as the modulus of the matrix falls.

Radiographs of some plates, in Fig. S7.7/2, give an impression of the through-thickness average of fibre orientation. When narrow beams are cut from various positions across the plate, the variation in their modulus reflects the variation in fibre alignment (or molecular alignment in unreinforced plastics). A test on an entire plate gives a mean value of the moduli of the individual beams and reduces the bias induced by special features such as the enhanced molecular and/or fibre alignment at an edge (edge effects contribute to the favourable alignment in end-gated bars). Similarly,

Single feed

'Coathanger' feed

Double feed

S7.7/2. Radiographs of differently gated square plaques. The patterns of fibre orientation correlate loosely with the measured anisotropy ratios (Table S7.7/1).

a combination of 0° and 90° data from a plate is superior to the single along-flow datum from a beam when an observed trend of modulus with respect to a structural variable is being explored. This is because the single datum cannot discriminate between a direct causal relationship and an indirect one, e.g. the influential factor may be a modified flow geometry which affects the 0° and 90° moduli in different senses.

Thus, the flexed square plate is a helpful technique, but it is nevertheless limited in that it gives the modulus only along the 0° and the 90° directions which is often insufficient to characterize the anisotropy. To avoid the cost and tedium of using narrow beam specimens cut from the plates at various

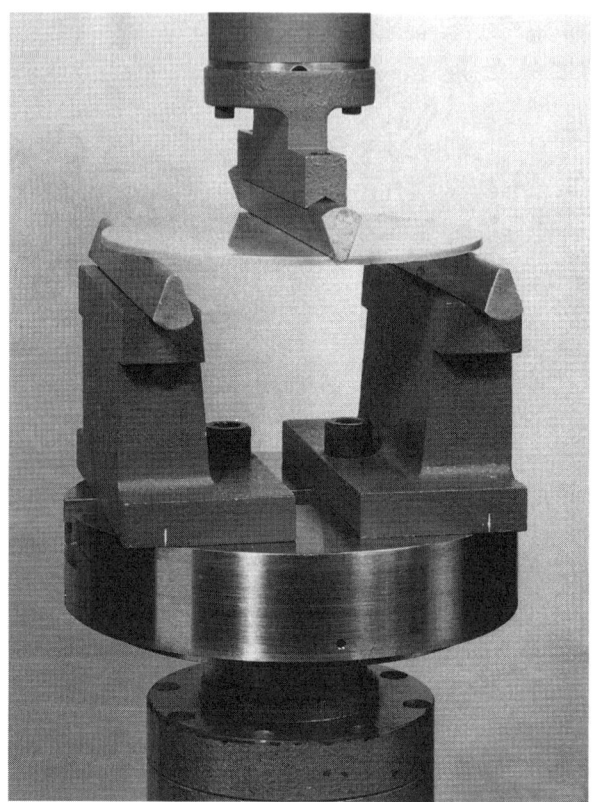

S7.7/3. Flexed disc subassembly. This test configuration enables the in-plane modulus anisotropy to be mapped comprehensively by a simple procedure.

angles to the principal axes, Stephenson *et al.*[1] used the novel means of a disc, rather than a square plate tested in three-line flexure, see Fig. S7.7/3. If the disc is rotated relative to the supports between successive ramp/reversed ramp excitations, the relative flexural anisotropy may be obtained as a function of angle of rotation. The raw data often suffice, but they can be converted if necessary into flexural stiffness by calculation or calibration against an isotropic specimen of similar size and known stiffness. The test disc may be cut from a larger plate or moulded directly as appropriate. Typical results, for discs moulded from three polyethersulfone compounds are plotted in Fig. S7.7/4, which is directly reproduced from a computer printout. The isotropy of the disc moulded from the unreinforced compound is characteristic of polyethersulphone and many other polymers in their glassy state but anisotropy ratios of up to 1.4 are common for crys-

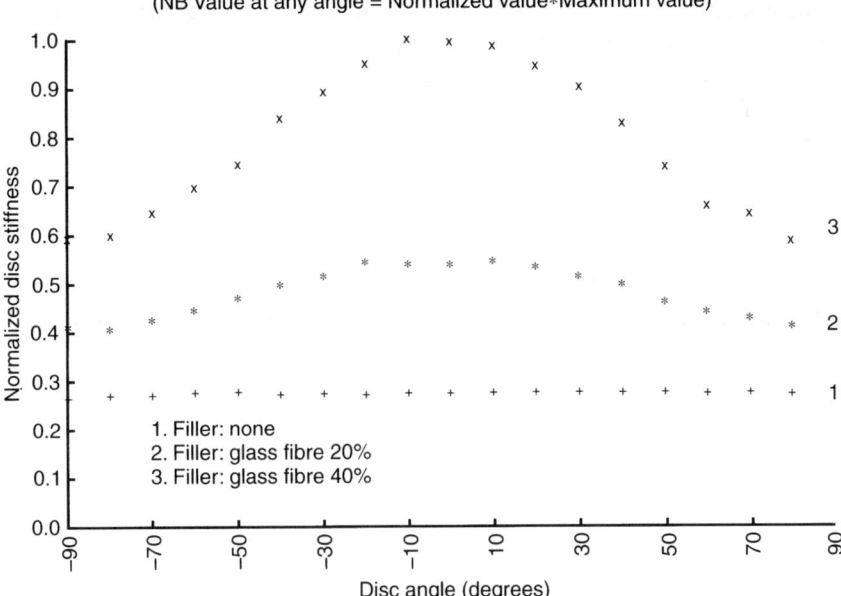

Maximum disc stiffness (GN/m^2) = 13.04
(NB Value at any angle = Normalized value∗Maximum value)

S7.7/4. In-plane modulus anisotropy of glass fibre reinforced poly(ethersulphone). (M. Whale, private communication.)

talline polymers (see Table 7.10/1, main text), and higher ratios for compounds containing fillers or short fibres.

It can be expected that the data produced by such methods are less accurate than those obtainable from conventional specimens. This may be due to the test pieces being not ideally flat and also because shear–tension coupling causes the discs to twist as they flex, but the sacrifice of accuracy and precision is offset by pertinence to end-products. The in-plane variation is a useful indicator of the degree of anisotropy that is likely to arise in panel mouldings and is invaluable to the experimenter when the influence of flow geometry or other manufacturing factors has to be either explored by, or eliminated from, an investigation. For instance, an observed correlation between tensile modulus as measured on an injection-moulded end-gated bar (which is the usual practice) and some structural change, say, to a plastic material may not be evidence of a direct cause–effect relationship. Instead, the structural changes may have affected the processability and hence the final anisotropy in the test bars, thereby distorting the trains of inference about a modulus-structure relationship. The flexed plate test would resolve the ambiguity.

A mean value for stiffness may suffice for many purposes. In principle, an edge-supported, centre-loaded disc should give a mean stiffness in a single measurement. However, that method tends to be unreliable because of the ill-defined stress state immediately under the loading nose and errors arising if the disc deforms asymmetrically. Nevertheless, a mean value can easily be derived from three-line flexure data such as those in Fig. S7.7/4 as the average of all the data, by visual estimate or as the mean value of an orthogonal pair of data. The average value for an orthotropic plate is given by the relationship:

$$8\bar{c} = 3c_{11} + 2c_{12} + 4c_{66} + 3c_{22} \qquad [S7.7.1]$$

where $\bar{c}$ = the mean plate stiffness.

Substitution into Eq. S7.7.1 of appropriate expressions for the isotropic case, i.e.

$$c_{11} = c_{22} = \frac{E}{(1-v^2)}$$

$$c_{12} = \frac{vE}{(1-v^2)}$$

$$c_{66} = G = \frac{E}{2(1+v)}$$

yields $\bar{c} = \dfrac{E}{(1-v^2)}$

Recent refinements in this area include the use of laminate theory to accommodate the change in the pattern of fibre orientation with depth below the surface of injection mouldings[2] and the use of computer simulation to predict it.[3]

Shear

Stephenson *et al.* also developed a version of the plate-twisting test. A square plate is subjected to the action of four point loads; two equal upward forces are applied at the ends of one diagonal and the same forces applied but downwards at the ends of the other diagonal; see Fig. S7.7/5. The effective plate shear creep modulus can be calculated from Eq. S7.7.2.

$$G = A\frac{L_d^2 P}{h^3\delta} \qquad \cdot [S7.7.2]$$

where L_d = diagonal span
$\quad\quad\quad P$ = applied force
$\quad\quad\quad h$ = plate thickness
$\quad\quad\quad \delta$ = relative movement of the loading point pairs
$\quad\quad\quad A$ = $\frac{3}{8}$

S7.7/5. Subassembly for the measurement of the shear modulus of plates.

Earlier versions of the plate torsion test either measured the central deflection relative to the corners or fixed three of the corners in a plane and measured the out-of-plane displacement of the other corner. The precision and accuracy of the method is not high because the test configuration verges on the unstable and because the plate tends to bend initially as it twists if it is not perfectly flat and because of tension–shear coupling. When the plate bends, the load–deflection curve is concave upwards rather than a straight line or convex upwards and the derived modulus apparently increases with the applied load. For example, the three plaques that are featured in Table S7.7/1 gave, respectively, the shear modulus values of 1.15 GPa^2, 1.15 GPa and 1.06 GPa at low loads (which differed because the thicknesses differed) and 1.28 GPa, 2.10 GPa and 1.16 GPa at doubled loads. The large disparity between the two results for the asymmetrical, single-feed plaque may have arisen because that plaque is more prone to postmoulding distortion than the other two.

Excessive local deformation (indentation) may occur if ideal 'point loading' is approached too closely; foam-core mouldings are prone to such indentation but special precautions are commonplace for many types of mechanical test on such structures.

Subsequently the method was adapted by Christie et al.[4] for a disc specimen, which allows shear modulus anisotropy to be measured along various directions in the plane, in the manner of Stephenson for flexure. See Chapter 3 Supplement S3.2.4 for more details.

References

1 R. C. Stephenson, S. Turner and M. Whale. 'The load-bearing capability of short-fiber thermoplastics composites – a new practical system of evaluation.' *Polymer Engineering and Science* 19, 3, 173, 1979.

2 D. R. Moore, J. M. Smith and S. Turner. 'Engineering design properties for injection moulding compounds based on poly(ether ether ketone).' *Plastics, Rubber and Composites: Processing and Applications* 21, 1, 19, 1994.

3 K.-C. Ho and M.-C. Jeng. 'Fibre orientation of short glass fibre reinforced polycarbonate composites under various injection moulding conditions.' *Plastics and Rubber Composites: Processing and Applications* 25, 10, 469, 1996.

4 M. A. Christie, M. W. Darlington, D. McCammond and G. R. Smith. 'Shear anisotropy of short glass fibre reinforced thermoplastic injection mouldings.' *Fibre Science and Technology* 12, 167, 1979.

8

Strength, ductility and toughness: general principles

8.1

If a material is brittle under a particular set of test conditions, the conventional tensile stress–strain curve, see Chapter 9, is approximately linear and ends abruptly as the specimen breaks. In such cases the strain and extension are small and hence the experimental curve of force vs. either time or displacement should be readily convertible into a stress vs. strain relationship. Such is the behaviour for many glassy thermoplastics and most thermoset materials. Even so, the accuracy of the calculated strain may be low because of the several sources of possible error that exist at low strains, as discussed elsewhere. Also, the apparent stress at failure is likely to be only an approximation to the true tensile strength because imperfections in the specimen and misalignment of the specimen–machine system induce premature failure. Misalignment causes inadvertent bending and hence varying stress over any cross-section.

In the past there was often an element of unpredictability about brittle fracture. Even now, with a satisfactory theory of fracture mechanics (see later in this chapter) and an understanding of how polymer architecture governs molecular mobility and the response of a test piece or other item to an excitation, brittleness still arises unexpectedly sometimes. A propensity to brittleness in a plastic item depends primarily on inherent factors such as the polymer architecture, molecular weight and molecular weight distribution of the polymer. In addition, it depends on the composition of the plastic, on geometric features such as corners, ribs, grooves and on adventitious factors such as inclusions, surface damage and degradation. In accordance with fracture mechanics principles, it also depends on the size of the item, large ones being more prone to brittleness than small ones made from the same material.

Anisotropy can also be detrimental. For many years, anisotropy of strength was acknowledged, almost surreptitiously, through recognition of a 'weak direction' in mouldings, extrudates etc., and assessed by the use of specimens whose main geometric axis had lain along the direction known, or thought, to be the weakest. That weakest direction is not readily identifiable in melt-processed items because of the complexity of the flow geometries that arise during the processing stage. The point-to-point variation of the tensile strength and the strength anisotropy in injection-moulded items has been well documented and is generally worse in grades reinforced with short glass fibres; the pattern of variation is consistent with the through-thickness average of the fibre orientation (and molecular orientation) pattern. In continuous fibre composites a 'weak direction' can often be avoided by judicious placement of the fibres.

8.2

If a material is neither classically brittle nor unambiguously ductile, the tensile force–displacement trace will curve over as the point of ultimate failure is approached and may even exhibit a maximum that is superficially indicative of yielding in the specimen.

The force will not translate accurately into a stress if the extension is substantial, unless a realistic correction can be applied to the cross-sectional area, or if an incipient neck entails a reduced cross-section. Similarly, the local strain at the failure site cannot be deduced from the overall extension or from a direct strain measurement at some other point. Apart from such macromechanical limitations, there are many forms of local damage, see Supplement S8.2, that affect the shape of the force–displacement curve, complicate the process of its translation into a stress–strain relationship and influence the nature of the final failure.

8.3

Many years ago, Bueche and White[1] observed that the maximum velocity of crack propagation in glass, v_c, is approximately equal to $0.5v_t$, where v_t is the velocity of propagation of transverse waves in an infinite medium, and that a wide range of plastics conform to the same rule. They inferred that the velocities were therefore independent of the type of bond that was broken in the fracture event.

$$v_t = [E/\rho(1+v)]^{1/2} \qquad\qquad [8.3.1]$$

where v = Poisson's ratio
         ρ = density

and hence $v_c \approx v_t/2 = (E/\rho)^{1/2}[8(1+v)]^{-1/2}$ $\qquad\qquad [8.3.2]$

The associated rate of strain at the crack tip is high which, on viscoelastic grounds, should favour brittleness but, on the other hand, in conjunction with the low thermal conductivity, may cause a rise in temperature locally and some ductility.

8.4

If a material is thoroughly ductile, the conventional tensile force–deformation curve exhibits a discontinuity in the slope that is related to yielding in the specimen. The force at yield can be measured accurately and a notional yield stress can be deduced; it represents the absolute limiting stress for most load-bearing applications. It is a viscoelastic quantity that decreases in magnitude with increasing temperature and with decreasing straining rate. It may be governed by the same molecular mobilities that govern the modulus behaviour but on a larger scale of movement, although it is often associated with breaking of secondary molecular bonds. Yield stress is widely disseminated as a function of temperature and straining rate.

For most thermoplastics, there are test conditions under which a tensile specimen first yields and necks and then extends non-uniformly to several times its original length before it breaks. The process is known as 'cold-drawing' and it has some practical relevance to film and fibre technology. The final form of the product is produced by a controlled drawing process where the local strains can be very high and the behaviour is governed by the postyield characteristics that are indicated by the shape of the force–deflection curve.

8.5

The theory that was developed to account for unreliable or inaccurate strength data for brittle materials and unexpectedly brittle failure of some items in service was based on the worst case, namely the detrimental effect of an infinitely sharp crack-like inclusion. The local stresses near the tip of a crack in an infinitely large plate stressed in the direction normal to the plane of the crack may be expressed as:

$$\sigma_L(r, \theta) = \frac{\sigma_R \sqrt{\pi a}}{\sqrt{2\pi r}} f(\theta) \qquad [8.5.1]$$

where
r = the distance from the crack tip
$f(\theta)$ = the angular distribution of the stresses
a = crack half-length
θ = the angle measured anticlockwise from the line of the crack.
σ_R = applied stress at a point remote from the crack (notch)
$\sigma_L(r, \theta)$ = local stress

Thus the stress at a point in the vicinity of the crack tip depends on its polar coordinates. See also Supplements S8.5.1 and S8.5.2.

For a crack aligned along the x-axis and a stress acting along the y-axis, Eq. 8.5.1 has the forms:

$$\sigma_{xx} = \frac{\sigma_R \sqrt{\pi a}}{\sqrt{2\pi r}} \cos\frac{\theta}{2}\left(1 - \sin\frac{\theta}{2}\sin\frac{3\theta}{2}\right) \qquad [8.5.2]$$

$$\sigma_{yy} = \frac{\sigma_R \sqrt{\pi a}}{\sqrt{2\pi r}} \cos\frac{\theta}{2}\left(1 + \sin\frac{\theta}{2}\sin\frac{3\theta}{2}\right) \qquad [8.5.3]$$

$$\sigma_m = 0 \quad \text{(plane stress)} \qquad [8.5.4]$$

$$\sigma_{zz} = v(\sigma_{xx} + \sigma_{yy}) \quad \text{(plane strain)} \qquad [8.5.5]$$

$$\tau_{xy} = \frac{\sigma_R \sqrt{\pi a}}{\sqrt{2\pi r}} \sin\frac{\theta}{2}\cos\frac{\theta}{2}\cos\frac{3\theta}{2} \qquad [8.5.6]$$

On the axis of the crack $f(\theta)$ has the value unity and σ_{xx} can be characterized by a single term, K_I, known as the opening mode stress field intensity factor and equal to $\sigma_R\sqrt{\pi a}$. The magnitude of τ_{xy} is zero along that axis.

An infinitely wide cracked plate in tension is a special case and the stress field intensity factor for other configurations is given by:

$$K_I = \sigma_R Y \sqrt{a} \qquad [8.5.7]$$

where σ_R = stress at a point remote from the crack (notch)
   a = crack length
   Y = f(a/W)
   W = specimen width

The quantity Y has been tabulated for many configurations (it has the value $\sqrt{\pi}$ for an infinitely wide plate).

The critical stress field intensity factor, K_c, is the value of K at which brittle failure develops and it is essentially a measure of the brittleness of a material. K_I describes and quantifies the stress field at the notch tip. The critical value, K_{Ic}, at which the crack supposedly becomes unstable and grows catastrophically is assumed to be a characterizing material property and is allegedly independent of the geometry and the loading configuration. The status of this theory has been thoroughly investigated for viscoelastic materials, like plastics, and its application has been well established.[2] For example, there are results reported where a common value of K_{Ic} is seen to be derived from different test geometries. The successful application of linear elastic fracture mechanics (LEFM) to plastics requires that the materials are homogenous and behave as pseudoelastic. Otherwise, other approaches or modifications to the theory are required. Modest excursions away from pseudoelastic behaviour can be accommodated by modifications to the theory, whilst large scale plasticity at the notch or crack tip requires the adoption of elastic–plastic theories.[3] Supplement S8.5 discusses some of the helpful adoptions of linear elastic theory in describing fracture toughness.

8.6

In principle, the concept of K_c (or its energy equivalent G_c, the critical value of energy release rate) (see Supplement S8.5) confers two outstanding advantages. By use of a dominating crack, the casual scatter in strength data due to poor surface finish, contamination, etc. should be eliminated. Moreover, because of the geometry independence of K_c, the data should be directly applicable to design calculations provided realistic assumptions about permissible crack lengths or likely flaw sizes can be made. In practice, the situation is not so satisfactory because the situation at the crack tip rarely corresponds exactly to what it should be for a classically brittle material. In theory (see Eq. 8.5.1), the stress tends to infinity as the distance from the crack tip decreases to zero but, in practice, the local stress field is often modified by plastic yielding. The locus of $\sigma = \sigma_y$ is approximately representable by a circle of radius r_p where:

$$r_p = \frac{1}{2\pi}\left(\frac{K_c}{\sigma_y}\right)^2 \qquad [8.6.1]$$

or by an equation of the same form but with a slightly different coefficient, depending on the details of the model, so that r_p may be regarded as the average extent of the yielded region. See also Supplements S8.5.2, S8.6.1 and S8.6.2. The plastic zone is not necessarily a continuum and it is generally thought to be a craze-like structure. Models of that region should therefore allow for strain-hardening in the fibrils that bridge the crazes because such fibril reinforcement will affect the energy absorption efficiency of the plastic zone and the ability of the original crack to propagate through and beyond the zone.

8.7

A localized plastic zone at the crack tip is thus beneficial in three ways.

1 It modifies the stress field, limiting the high local stresses and thereby reduces the probability that the crack will grow.

2 It absorbs more energy than would be required for the equivalent growth of a brittle crack.

3 It dissipates energy in that the energy absorbed in a plastic zone is not recoverable and not available, therefore, to support further crack growth in the manner of the energy associated with elastic deformation.

Thus a straightforward argument links the size of the plastic zone with the ductility and the view has emerged that, in materials development programmes, the targets for the mechanical properties should be such that the plastic zone at the crack tip is as large as possible. That can be achieved either with a high value of K_{Ic} or with a low value of σ_y. A higher K_{Ic} can be attained in thermoplastics by a higher molecular weight, but that entails a higher melt viscosity and possible attendant processing difficulties; lower σ_y can be obtained by the incorporation of a rubbery phase, for instance, but that entails lowered modulus, see Supplement S8.7. Included rubber particles act in various ways, i.e. as crack stoppers, craze initiators, shear yield initiators, as do high modulus filler particles when they decouple from the matrix with increasing tensile strain. Strongly bonded high modulus filler particles tend to inhibit shear yielding and thereby reduce the fracture toughness, see Supplement S8.7. The issue of plastic zone size is more complex for thermosetting materials because the cross-linking inhibits yielding, but toughening can be achieved via flexible segments in the polymer chains and rubbery additives. In polymer–fibre composites local debonding creates pseudoplastic zones which increase the toughness, but the deliberate reduction of interphase bonding reduces the overall stiffening effect of the fibres, following the general rule that toughness and stiffness are reciprocally related.

8.8

The interrelationship between K_{Ic} and σ_y is useful in qualitative assessments during polymer development programmes. For example, the fracture toughness parameter of a propylene–ethylene copolymer is nearly independent of ethylene content but yield stress changes markedly, giving larger plastic zone sizes and greater toughness to lower temperatures. That rationalizes the greater toughness of the copolymers and justifies the use of simpler test procedures, e.g. impact tests at low temperatures, to monitor development progress.

8.9

Tests based on linear elastic fracture mechanics (LEFM) theory are now widely used for the study and assessment of the strength and toughness of brittle and marginally brittle plastics and composites, despite serious limitations to the validity of the theory in relation to ductile and/or non-linear viscoelastic systems. Several fracture testing protocols that define the current best practices have been developed by a European task group.[3,4] Some of the common test configurations are described in Chapter 9 (that arrangement was chosen partly for convenience and partly because fracture mechanics investigations are commonly based on ramp excitation tests but any of the basic excitations may be applied to fracture mechanics configurations).

The theory has now been extended beyond the linear elastic regime and some early experimentation has been developed.[3] Thus, the test procedures for plastics etc., use the same configurations as those for metals but with lowered temperatures and higher deformation rates used to limit ductility as necessary.

8.10

The tests for Mode I fracture toughness use sharply notched (cracked) specimens, either in otherwise conventional configurations, i.e. tensile bars, beams in flexure, or in special configurations such as the compact tension specimen or the double cantilever beam, see Chapter 9, Supplement S9.8.2. There are two principal experimental difficulties. One is the choice of appropriate specimen shape and dimensions to ensure that the crack grows in conformity with the idealizations of the theory. The other is the identification of the critical point on the force–displacement curve, which is not necessarily associated with the peak force or with any particular subsidiary feature on the curve. At slow straining rates visual observation of the moving notch tip may provide the corroborative evidence, but for high straining rates photography or video recording may be required.

In relation to the choice of specimen, the main experimental precaution is that plasticity at the crack tip should be limited, e.g. a lowered test temperature raises the yield stress and thereby reduces the size of the plastic zone; specimen dimensions that are large relative to the plastic zone favour plane strain conditions. In that case, LEFM can be used, with the crack length, a, modified by a term related to the size of the plastic zone. Thus:

$$a_f = a + r_p \qquad\qquad [8.10.1]$$

where a_f = effective crack length
       a = crack length
      r_p = length of line plastic zone (Dugdale model)

The Dugdale line plastic zone is deemed to be the most appropriate for that purpose.

$$r_p = \frac{\pi G_C}{16 w_p} = \frac{\pi}{16}\left(\frac{K_C}{\sigma_y}\right)^2 \qquad\qquad [8.10.2]$$

where G_C = critical strain energy release rate
      w_p = energy per unit volume to yield
      K_C = critical value of stress field intensity factor
      σ_y = tensile yield strength

The terms K_C and G_C are seldom referred to by their formal definitions and nowadays are both called fracture toughness.

8.11

Localized damage as described in Section 8.2 and Supplement S8.2 may be imperceptible or unobservable under visual scrutiny. Such damage or incipient damage in an end-product can be detrimental to its serviceability and early detection *in situ* may be vital as, for example, in structural composites. Non-destructive testing of composite structures in the aircraft and associated industries is mentioned briefly in Supplement S8.11. There are also tests that assess the detrimental effects of prior impact and/or definite impact damage on other mechanical properties.

8.12

The yield stress is essentially a measure of the ductility of a material; beyond that point a test specimen (or a service item) either fails by 'necking rupture' or suffers gross plastic distortion. The degree to which rupture is delayed as the material yields progressively depends on its propensity to work harden. Dominant work hardening arises if adequate molecular alignment is established in the neck; a high molecular weight and molecular entanglements are favourable in that respect.

The von Mises' yield criterion was, at one time, regarded as a reasonable representation of the behaviour of isotropic samples of ductile plastics, i.e. shear yielding would occur when

$$6J_{2D} = (\sigma_1 - \sigma_2)^2 + (\sigma_2 - \sigma_3)^2 + (\sigma_3 - \sigma_1)^2 \qquad [8.12.1]$$

where σ_1, σ_2 and σ_3 are the three principal stresses and $6J_{2D}$ is the second invariant of the deviatoric part of the stress tensor.

However, Eq. 8.12.1 is potentially unsatisfactory for a plastics material because the retardation spectra for shear deformations are different from those for the bulk deformations. It is now often assumed that yielding will occur in tension when:

$$\sigma_y = AJ_{2D}^{\frac{1}{2}} + BJ_1 \qquad [8.12.2]$$

where BJ_1 is the first invariant of the stress tensor and A and B are coefficients that can be functions of temperature and straining rate. Development of both the theory and the experimental verification pose difficulties that are not yet fully resolved but see reference 5, for instance, for details.

8.13

End-products, including test specimens, are generally anisotropic with respect to strength and toughness. One cannot correctly state that the strength of a plastic is anisotropic, only that the strength as manifest in an item moulded from that plastic is anisotropic. The degree of anisotropy of strength is similar to that of modulus but greater sometimes because of a fundamental difference between the two groups of phenomena. The overall stiffness of a structure is some weighted mean of the separate stiffnesses of the component sections and, on balance, the stiffer elements support the less stiff ones. There is no comparable 'overall strength'; a failure initiates when the magnitude of the stress exceeds the strength at some point, the local stress field changes in consequence of the developing failure and any weakness is automatically exploited. Consequently, strength data tend to vary more with direction and position than modulus data do. That is certainly the case in the fracture behaviour for injection moulded discontinuous fibre reinforced plastics. Supplement S8.13.1 describes recent attempts to apply the concepts of LEFM to such materials. The approach can be only pragmatic because the materials are inhomogeneous and end-products are often anisotropic. On the other hand, application of LEFM to unidirectional continuous fibre reinforced plastics in order to measure delamination toughness is more rigorously aligned to LEFM theory as shown in Supplement S8.13.2, since the local stress field in the vicinity of the crack is operating in a less heterogenous environment.

Unfavourable molecular alignment affects the fracture toughness in two ways: it lowers the resistance to crack propagation by reducing the number of molecular chains that have to be broken by the advancing crack and it provides 'molecular notches', which can be spectacularly detrimental in some instances. Examples of the effect of molecular notches include fibrillation in highly forward-drawn film (which can be commercially useful) and unexpected brittleness of certain injection mouldings from which a very thin 'flash' has not been completely removed (the molecules tend to be very highly aligned there).

8.14

Apart from the anisotropy, strength data are more variable than modulus data in a statistical sense, partly because they are more sensitive to the inherent defects in a moulding and partly because measurements of strength are prone to greater errors. The combined effects of anisotropy and casual variability pose difficulties for both data generators and data utilizers, because test methods need to be fit for their particular purpose as do the data.

The standard tabulations of strength data are not very helpful since they generally give upper bound values derived from short-term tests on specimens with favourable molecular alignments. Recent developments in this field, largely in response to downstream discontent with the disparities between published data and service performance, should improve the situation but, as discussed elsewhere in relation to other mechanical properties, progress is generally slow when an international consensus has to be sought. Matters relating to the development of user-friendly databases on the strength of plastics are discussed in Supplement S8.14.

Until more definitive data are widely available designers should bear in mind that the combination of adverse flow geometry, a sharp notch and a long period under stress (or alternatively a fluctuating stress) can reduce the in-service strength of thermoplastics to one tenth of the published short-term strength. Similar deficiencies in strength can arise in some fibre-reinforced thermosetting resin systems when inadequacies in the processing stage cause resin starvation, unwetted fibres, fibre agglomerates, etc.

References

1 M. Bueche and A. V. White. 'Kinematographic study of tensile fracture in polymers.' *Journal of Applied Physics*, 27, 980, 1956.
2 J. G. Williams. *Fracture Mechanics of Polymers*. Ellis Horwood, Chichester, 1984.
3 D. R. Moore. 'Fracture of polymers and composites into the 21st century.' *Materials World* 1, 5, 272, 1993.
4 J. G. Williams and M. J. Cawood. 'European Group on fracture: K_c and G_c methods for polymers.' *Polymer Testing* 9, 15, 1990.
5 B. E. Read, G. D. Dean and B. C. Duncan. 'Characterisation of the strain rate and temperature dependence of the properties of plastics for the prediction of impact performance.' NPL Report CMMT(A)144, February 1999.

S8.2 Types of local deformation/damage in thermoplastics

Deformation damage at, and beyond, the yield point takes various forms, depending on the several influential factors cited elsewhere in this monograph that affect mechanical properties, i.e. polymer type, state of molecular order, composition and test or service conditions. In some materials, for example, biaxially drawn polyethylene terephthalate, the deformation is relatively homogeneous, in the sense that sudden discontinuities are not normally seen with an optical microscope. However, in many plastics specimens, discontinuous, inhomogeneous and localized deformation can be seen with an optical microscope and, in some cases, with the naked eye. The discontinuities take various forms, viz. plastic zones, crazes, deformation bands, stress whitening, microfibrillation. Much of the following information was originally collated by P. I. Vincent.[1]

Plastic zones

These form at the tips of cracks and notches. The shape of the zone varies from material to material, with the state of molecular order in the test piece and with the test conditions. The main classes of plastic zone are:

Kidney-shaped zones

Linear elastic theory predicts kidney-shaped strain contours around notches in stressed specimens and they are observed at low strains in many photo-elastic materials. They were also seen, even at high strains, in biaxially drawn polyethylene terephthalate and low density polyethylene.

Wedge-shapes zones

Long narrow zones, lying perpendicular to the applied stress, continuing the crack, have been seen in many polymers including isotropic polyethylene terephthalate and polycarbonate films.

Internal kidneys

In isotropic polyethylene terephthalate and polycarbonate films, with further straining, the wedge-shaped zone develops into a comparatively large area of highly oriented polymer; within that area a kidney-shaped zone appears round the crack tip.

Crazes

In polymethylmethacrylate and polypropylene the wedge-shaped zone may have a refractive index, and therefore density, considerably below that of the starting material; it is then reasonable to consider that the plastic zone is a single craze. Multiple crazing and irregular zones have been seen in polystyrene and rubber-modified polystyrene, polypropylene and other polymers. As polymers become more brittle, the amount of crazing diminishes towards a single craze.

Crazes can be seen in all non-crystalline polymers and in some crystalline polymers, polypropylene, methyl pentene polymers and polyethylene, but they have not been reported in polyamides.

Deformation bands

These are bands of discontinuous deformation higher than in adjacent material. Two types have been seen. Normal deformation bands lie perpendicular to the applied stress. Shear deformation bands lie at an angle between about 30° and 60° to the applied stress. They have been seen in PVC, PMMA and polycarbonate, among others. They appear to be precursors of necking and ductile failure, rather than of brittle fracture.

When strips of polycarbonate film are stretched in air, isochromatic fringe patterns indicate stress concentrations at edge defects. The lobes of these patterns gradually elongate with increasing strain. Shear bands form in the directions of the lobes eventually giving kink bands and complex cross-patterns. When strips are stretched with a small pool of ethanol on the surface, normal deformation bands appear, widen and merge to give complex patterns, quite different from those seen in air.

Normal deformation bands have been seen in rubber-modified polystyrene film, both notched and un-notched, forming between crazes. They have also been seen in polycarbonate film stretched in ethanol.

Stress whitening

Many plastics become white and opaque when stressed in tension close to the yield point. It is particularly common in heterogeneous materials and is attributable to phase separation; it is also common in crystalline polymers, which may be regarded as heterogeneous at the level of crystal and spherulite sizes. The degree of stress-whitening in PVC, for instance, depends on its compositional formulation. It can be intense with particulate additives, but practically non-existent with soluble additives. The density falls as the specimen whitens but not by more than 5%. Transmission interferometry on thin sections of stress-whitened PVC shows that it is patchily variable in refractive index on the scale of 1–3 mm. This is assumed to correspond to variations in density and plastic strain, and may account for the light scattering giving rise to the whiteness. Control specimens and plastically deformed but non-whitened specimens do not show the same patchy variability. Transmission electron microscopy on ultramicrotome sections from stress-whitened regions shows no signs of holes or crazes at least down to the 50 Å scale. Stress whitening evidence of phase separation is often seen in fibre composite materials.

Microfibrillation

Fibrous textures, on a wide range of scales, are often seen in optical and electron micrographs of deformed or broken specimens.

It is not always obvious from inspection by the naked eye or simple microscopy whether a given defect is a craze, a crack or a normal deformation band. Transmission interferometry on thin sections cut across the defect provides a good means of making this distinction. Measurement of the refractive index is helpful; if the refractive index at the defect is equal to that of the immersion fluid the defect is probably a crack; if it is intermediate between that of the fluid and that of the bulk plastic, the defect is probably a craze.

The classes of defect or damage that are described above are not specific to particular straining rates but they may be specific to particular plastics, particular deformation modes and/or particular levels of anisotropy. They also often occur in combination, especially so in fibre composites, where heterogeneity and anisotropy interact in complex ways. Thus, in rubber-toughened plastics voiding in the matrix is often superposed on the shear and craze deformation processes.[2] Shear deformations are assumed to occur at constant volume; crazes and voids cause the overall volume to increase, which can be detected by creep tests or conventional tensile tests during which the lateral contraction is measured simultaneously.

The various forms of deformation damage described above may be regarded as prefailure phenomena which, depending on circumstances, are not necessarily critically limiting when they arise in service, but they may become initiation sites for brittle failure. When that occurs the prefailure damage phenomena tend to distort subsequent fracture mechanics calculations.

Reference

1 P. I. Vincent. Private communication.
2 F. Ramsteiner. 'Structural changes during the deformation of thermoplastics in relation to impact resistance.' *Polymer*, 20, 839, 1979.

S8.5   Fracture toughness

S8.5.1 Stress concentration and stress field intensity

The stress concentration factor, k, is defined as the ratio of the maximum stress at the root of a notch to the nominal applied stress. For an elliptical notch:

$$k = \frac{\sigma_{max}}{\sigma_{nom}} = 1 + 2\sqrt{\frac{a}{\rho}}$$ [S8.5.1.1]

where a = notch depth
ρ = notch-tip radius

Thus, for a circular hole, $a = \rho$ and $k = 3$, and for sharp cracks, $\rho \to 0$, so that:

$$\sigma_{max} \approx 2\sigma_{nom} 2\sqrt{\frac{a}{\rho}}$$ [S8.5.1.2]

and the crack-tip stress approaches infinity. However, the product of the maximum crack-tip stress and the square root of the radius of the crack-tip as that tends to zero remains finite:

i.e. $\sigma_{max}\sqrt{\rho} \approx \sigma_{nom} 2\sqrt{a}$ [S8.5.1.3]

and $\dfrac{\sigma_{max}}{2}\sqrt{\pi\rho} \approx \sigma_{nom}\sqrt{a\pi}$ [S8.5.1.4]

which conforms to the fracture mechanics solution for an infinitely wide plate.

S8.5.2 The fracture toughness parameters

Fracture toughness is a measure of the resistance to fracture of a cracked specimen, i.e. a specimen with a very sharp notch. It may be expressed either as the critical value of a 'stress field intensity factor' (K_c), see Section 8.5 or as the energy required to create unit area of new cracked surface (G_c). The underlying theory for the latter developed from the basic concept of Griffith that the strength of a brittle material is governed by small defects that act as stress concentrators and, if a dominating crack is introduced, measurement of the strength should allow the fracture surface energy to be calculated.

It transpires that experimentally derived values of G_c are in the region of 0.3 kJ/m^2 for a brittle polymer and are several orders of magnitude greater than values calculated from molecular constants. The explanation for that disparity is that fracture surfaces are seldom ideally brittle; the experimental fracture surface energy is mainly the energy to yield of a thin surface layer that was created by small-scale yielding at the tip of the advancing crack. Leaving such molecular niceties aside, the breaking stress for an infinitely wide sheet containing a crack of length $2a$ lying orthogonally to the stress axis is given by

$$\sigma_f = \left(\frac{2E\gamma}{\pi a}\right)^{1/2}$$
[S8.5.2.1]

where
σ_f = tensile strength
E = tensile modulus
γ = fracture surface energy
a = half-length of intrinsic defect

The energy to propagate the crack comes from the elastic stored energy and

$$\frac{dU}{da} = 2\gamma = G$$

where G = strain energy release rate
whence

$$\sigma_f = \left(\frac{GE}{\pi a}\right)^{1/2}$$
[S8.5.2.2]

From the alternative starting point, a uniaxial stress σ acting on an infinite plane sheet bearing a crack of length $2a$ lying perpendicular to the stress axis gives rise to a 'crack opening mode stress field intensity factor', K_I, given by the equation.

$$K_1 = \sigma\sqrt{\pi a}$$
[S8.5.2.3]

K_I increases with σ or a until a critical value, K_{Ic}, is reached, at which point the crack rapidly progresses to total failure. It follows that:

$$\sigma_f = \frac{K_{Ic}}{\sqrt{\pi a}}$$
[S8.5.2.4]

and since the strain energy release rate has a critical value, G_{Ic}

$$K_{Ic}^2 = EG_{Ic}$$
[S8.5.2.5]

Equations S8.5.2.2 and S8.5.2.4 summarize the requirements for toughness, namely a high fracture toughness and/or a small defect size. The former depends on the length of the molecular chains, the degree of entanglement and other structural features; the latter depends on the scale of the molecular and other inhomogeneities, the degree of molecular order and the incidence of flow defects.

There are three modes of crack propagation, viz.

1   Opening mode, K_I.
2   Shear mode, K_{II}.
3   Antiplane shear mode or tearing mode, K_{III}.

The concept of K_{Ic} is a simplification in that the state of stress at the crack tip changes through the thickness. A state of plane stress exists at the surface, which allows plasticity to develop, whereas a state of plane strain exists in the interior of thick sections, which restricts plasticity. Thus, there is a distinct thickness effect which sometimes entails two fracture toughness quantities, namely,

1   K_{C_1} plane strain value of stress field intensity factor
2   K_{C_2} plane stress value of stress field intensity factor

On all occasions, it can be expected that $K_{C_1} < K_{C_2}$. Consequently, the plane strain value of stress field intensity factor (K_{C_1}) is a minimal value in analysing brittle fractures.

To allow for limited ductility or plasticity around the crack tip, fracture mechanics theory postulates a plastic zone, bounded by the limit of yielding, of magnitude (plane stress case):

$$ r_P = \frac{1}{2\pi}\left(\frac{K_{Ic}}{\sigma_y}\right)^2 = \frac{1}{2\pi}\left(\frac{EG_{Ic}}{\sigma_y^2}\right) \qquad \text{[S8.5.2.6]} $$

where σ_y = tensile yield stress

in the fracture direction for plane stress. The analysis, based on stress distribution and an appropriate yield criterion, predicts a zone which extends a comparable distance normal to the plane of fracture.

A planar plastic zone which could be regarded as a model craze, was proposed by Dugdale. The analysis in this case, again for plane stress, gives a line length zone of:

$$ r_p = \frac{\pi}{16}\left(\frac{K_{Ic}}{\sigma_y}\right)^2 \qquad \text{[S8.5.2.7]} $$

and crack opening displacement:

$$ \text{COD} = \delta = \frac{8\sigma_y a}{\pi E}\ln\left[\sec\frac{\pi\sigma_f}{2\sigma_y}\right] \qquad \text{[S8.5.2.8]} $$

Fracture is assumed to occur when δ reaches a critical value δ_{crit}.
For $\sigma_f \ll \sigma_y$, Eq. S8.5.2.8 simplifies to:

$$ \delta = \frac{G_{Ic}}{\sigma_y} = \frac{K_{Ic}^2}{\sigma_y E} \qquad \text{[S8.5.2.9]} $$

The plastic zone size, r_p, of Eq. S8.5.2.6 can then be written as:

$$r_p = \frac{1}{2\pi}\left(\frac{E}{\sigma_y}\right)\delta \qquad\qquad [S8.5.2.10]$$

i.e. the crack opening displacement sets a limiting value for r_p and also

$$G_{Ic} = \sigma_y\delta \qquad\qquad [S8.5.2.11]$$

which can be interpreted as indicating that brittle fracture is merely a special case of ductile failure.

It had been postulated for metals that a given value of δ needs to be accommodated by a specific plastic zone size. Then, for a particular crack and a critical value of δ the issue of whether brittle fracture or general yielding occurs depends on the position of the boundary relative to the crack. If the plastic zone traverses the net cross-section before δ reaches its critical value, the failure is ductile, but it will be brittle if the reverse is true, which may occur if the boundary is remote, i.e. the size effect in a slightly different guise.

Where there is limited ductility at the crack tip the effective crack length can be taken as $a + r_p$. The fracture mechanics analysis is reliable only if the path to be traversed by the growing crack is much greater than $4r_p$. If the path length is less than $4r_p$, the fracture is likely to be ductile and linear elastic fracture mechanics analysis is then generally inapplicable, although elastic plastic fracture mechanics through either the determination of the essential work of fracture or a crack resistance parameter (J_c) might be helpful.[1,2]

There is another constraint on the permitted size of a plastic zone for plane strain conditions to prevail at the crack tip, in order for the measured stress field intensity factor or the strain energy release rate to be valid in an LEFM sense. The thickness of the specimen must be large relative to the size of the plastic zone. If the thickness is similar to the zone length, unrestrained yielding occurs over the cross-section. A generally accepted criterion is:

$$\text{Width} \geq 2.5\left(\frac{K_c}{\sigma_y}\right)^2$$

Additionally, r_p should be much smaller than the crack length, by the same inequality relationship. Undoubtedly, the plastic zone is an anomaly in linear elastic fracture mechanics analysis.

Toughness can be approached experimentally through these linear elastic fracture mechanics parameters; it can also be approached by the more traditional strength data. In making a selection of methodology for assessing toughness or strength, a number of questions arise:

1 Is it necessary to measure a formal fracture toughness property or would a more arbitrary measure suffice?
2 How effective are the data likely to be in the context of the investigation/evaluation?
3 How reproducible are the prospective fracture toughness data?
4 Are there severe experimental difficulties in obtaining rigorous fracture mechanics parameters and, if so, what are the advantages in a pragmatic adoption of such parameters over the alternative approaches?

Clearly, the decision on the test tactics must depend on the ultimate purpose of the test programme, a cost–benefit analysis and the likely penalties attendant upon a misleading outcome.

References

1 Y. W. Mai and B. Cotterrell. 'The essential work of fracture for shearing of ductile materials.' *International Journal of Fracture*, 24, 229, 1984.
2 S. Hashemi and J. G. Williams. 'Fracture characterization of tough polymers using the J method.' *Plastics and Rubber: Processing and Applications* 6, 4, 363, 1986.

S8.6   Crack-tip plastic zones and ductility

S8.6.1 Ductility factor

In principle, the shape and size of the plastic zone can be calculated from Eq. S8.5.2.6–8 and a yield criterion, but the latter is a poorly defined quantity, (Section 8.12) and the computed shape is likely to be an approximation. Even so, the relationships Equations S8.5.2.6 and S8.5.2.7 suggest that the quantity $\left(\dfrac{K_{Ic}}{\sigma_y}\right)^2$ governs plastic zone size (and hence ductility) and therefore 'ductility factor' has been defined as:

$$D = \left(\frac{K_{Ic}}{\sigma_y}\right)^2 \qquad\qquad [S8.6.1.1]$$

There is, however, an intellectual difficulty, namely it is unreasonable to expect that a test piece can be simultaneously ductile and brittle under one particular set of test conditions. The pragmatic solution is to measure K_{Ic} in a brittle region and σ_y in a ductile region, and to extrapolate the separate data to a common test condition or adjust them by an independent rationale. This has been done for Table S8.6.1/1, where the values of D relate to an elapsed time of 10^5 seconds; σ_y was the stress for ductile failure from a creep rupture test (see Chapter 10) and K_{Ic} was the critical stress field intensity factor for brittle failure in a fatigue test (see Chapter 10 also).

The values of ductility factor in Table S8.6.1/1 give credible comparisons of the materials. Plastic zone size, Eq. S8.6.1.1, gives the same comparison

Table S8.6.1/1 Experimental values of ductility factor

	σ_y (10^5 s) (MPa)	K_{Ic} (fatigue failure at 10^5 s) (MPa m$^{1/2}$)	D Ductility factor (mm)
Unfilled nylon 66	73	0.62	0.07
30% glass-filled nylon	84	1.79	0.45
PMMA	51	0.39	0.06
UPVC	39	0.48	0.15
PES	70	0.63	0.08

Specimen states for the creep rupture (σ_y) and the cyclic fatigue (K_{Ic}) were the same.

of relative toughnesses, of course. However, those particular types of data entail careful specimen preparation and long experimentation times; shorter and simpler tests giving analogous information would be preferable. A possible alternative is the comparison of the strengths of notched and unnotched specimens but with some manipulation of the data, see Supplement S8.6.2.

Ductility factor, in common with the other fracture toughness data, is a material property relating to ductility. The toughness of a product entails the ductility of the material and the geometry and size of the item. If the size of the notional plastic zone is small relative to the dimensions of the item, failure may be brittle even though the material is ductile. By a reverse argument, there are constraints on the recommended dimensions of fracture toughness test pieces, see Supplement S8.6.2.

S8.6.2 Sharp notches (cracks)

As a general rule of thumb, the notch (crack)-tip radius should be an order of magnitude smaller than the plane strain value of plastic zone size (r_p):

$$r_p = \frac{1}{6\pi}\left(\frac{K_c}{\sigma_y}\right)^2 \qquad\qquad [S8.6.2.1]$$

With materials of low fracture toughness, fallacious materials comparisons may be made if this qualification is not satisfied, as shown in the following example.[1]

Figure S8.6.2/1 shows fracture toughness vs. glass fibre content for some toughened and untoughened phenolic composites (discontinuous fibre reinforcement). It clearly shows that the toughened compounds, as expected, exhibit higher fracture toughness. These measurements were based on application of LEFM where all the necessary requirements for achieving valid fracture data were satisfied.

When the same materials are evaluated by a standard notched Izod pendulum impact test, where the notch-tip radius is 250 microns, there is no difference in impact performance between the toughened and untoughened compounds. This is shown in Fig. S8.6.2/2.

From measurements of K_c and yield strength for these compounds, it is possible to calculate the notional plane strain plastic zone size and to determine the ratio of notch tip radius to plastic zone size. This ratio should be less than 0.1 if the notch is sharp enough to act as a natural crack. For the notched Izod results that ratio varies between 1000 and 10 (unreinforced resins to 60% w/w glass-reinforced compounds, respectively). Therefore, crack blunting occurs in the Izod tests and provides misleading results for the measured toughness.

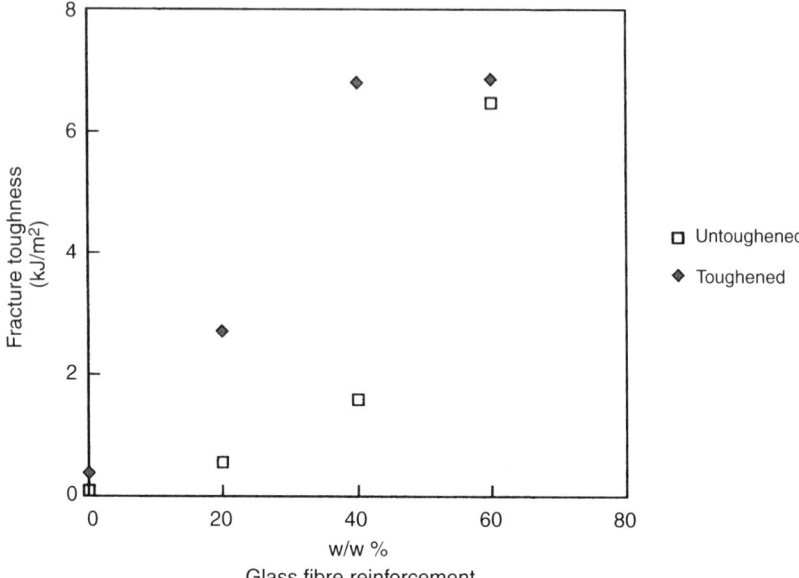

S8.6.2/1. Fracture toughness (G_c) vs. glass fibre content for some phenolic compounds[1].

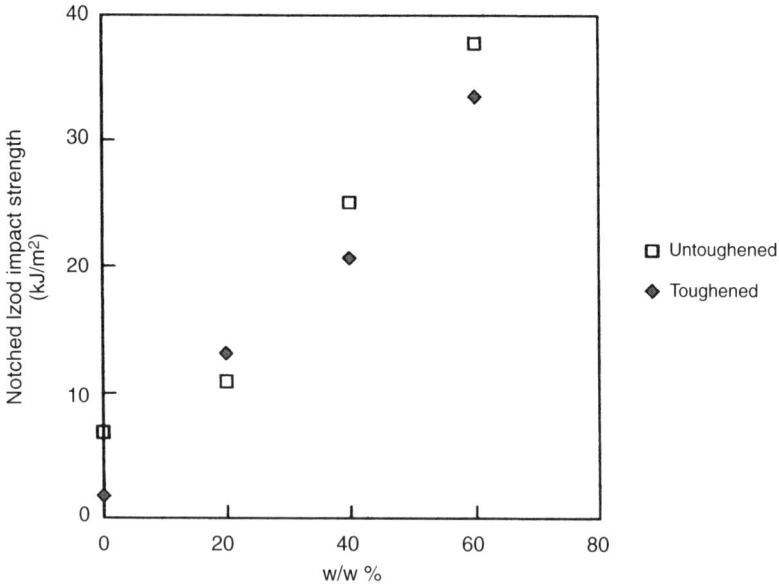

S8.6.2/2. Notched Izod impact strength vs. glass fibre content for some phenolic compounds as those shown in Fig. S8.6.2/1[1].

One can regard a notch-tip plastic zone as effectively blunting what would otherwise be a sharp crack. In a reverse argument it follows that the outcome of a fracture toughness test may depend on the 'quality' of the starter crack. Sharp notches are difficult to produce in materials in their ductile state, which tend to yield under the notching tool. Some brittle materials, e.g. polystyrene, are troublesome also, because they craze ahead of the notching blade and the crazes constitute a pseudoplastic zone. Notching at lowered temperature and/or by means of very sharp cutting edges subjected to sharp blows is sometimes helpful, provided the material is not prone to craze or to phase separation. Another technique is to generate a sharp crack from a blunter notch by intermittent stressing, which sometimes promotes brittleness, and then to create a new specimen incorporating the crack.[2]

A very sharp notch may not be necessary for tests in flexure because a surface notch has a high constraint in that configuration. Similarly, if a crack and its plastic zone are small relative to the size of the stressed item, then the crack may grow catastrophically irrespective of the actual sharpness of the notch. This is manifest dramatically by occasional unexpected brittle failures in service items manufactured from materials generally regarded as ductile.

References

1 A. C. Lowe, D. R. Moore and P. M. Rutter. *Impact and Dynamic Fracture of Polymers and Composites*, Ed. J. G. Williams and A. Pavan. ESIS Pub 19, MEP (London), 1995, pp. 383–400.
2 J. S. Foot and I. M. Ward. 'Fracture behaviour of PET.' *Journal of Materials Science*, 7, 367, 1972.

S8.7 Ductility and toughness in filled plastics

A significant fraction of the total effort directed towards the development of new plastics or enhanced variants of old ones is expended on a search for exceptions to the modulus/ductility reciprocity rule. The formulations and the processes by which they are created are often complex. However, stripped of the sophisticated details, they generally reduce to the incorporation of a toughening agent, often a polymer that is rubbery at the anticipated service temperature, with possibly a third high modulus phase such as inorganic particles to compensate for the drop in modulus caused by the rubber. The advocates of any particular formulation can normally cite some apparent advantage as indicated by one or more of the standard tests, but such supporting evidence is usually valid in only a narrow context. Thus, for example, blends that have been developed for enhanced impact resistance may have inferior creep rupture or fatigue lifetimes. Also, in a filled plastic with poor coupling between filler and matrix, the yield stress may be lowered by 'false yield decoupling'. This occurs without the low-strain modulus being adversely affected, but that seemingly favourable improvement in the modulus/ductility balance may be an illusion because K_c may also be lowered in the false-yield zone (because the material there is essentially a foam).

The property advantages conferred by the incorporation of fillers may be offset by the higher production costs (raw material and end-product). The additive has to be uniformly distributed throughout the mass of the material. This entails an extra stage in the processing operations and, if the volume fraction of the second phase is greater than about 0.10, there is usually the penalty of impaired processability, which increases the operational costs through increased power consumption and longer cycle times.

The greater ductility of a rubber-modified grade arises partly from the low temperature loss process of the rubbery phase and therefore an energy absorption capability. The associated influential mechanisms may involve cavitation in the rubber, crazing in the matrix or between a rubber particle and the matrix, debonding around a particle, and the blunting of crack tips by the dispersed rubbery particles.

High modulus fillers increase the modulus and sometimes increase the strength, but they almost invariably reduce the ductility. That deficiency is a direct consequence of the heterogeneity, the interfaces and agglomerates of poorly dispersed filler being initiation sites for cracks. Thus, phase heterogeneity enhances or reduces ductility depending on whether the included phase is deformable or rigid; presumably stresses at the interfaces are relieved by strains in the rubbery particles whereas there is no such mechanism possible with rigid particles and strain accommodation occurs via phase separation.

There is much more to the technology of toughening or stiffening additives than is covered by those simple statements, but the basic principle is nevertheless the utilization of heterogeneity to achieve particular objectives. Thus, when the interface is influential, the size of the included particles is likely to be important because, for a constant volume fraction the surface area is proportional to $1/r$ for spheres. On the other hand, a practical lower limit on particle size is reached because small ones are difficult to wet and to disperse; agglomerates reduce the effective interface area and also act as flaws. As a general rule, in contrast to ductility, modulus is virtually unaffected by 'grain size' and other details of the heterogeneity. This is probably because the modulus of a composite specimen is a relatively straightforward weighted mean of the moduli of the constituent phases, provided the 'grain size' is small relative to the size of the test item. However, fracture is a local event that progresses from one volume element to another and the associated strength and ductility are therefore susceptible to variations in the structural texture.

The history of the development of toughened or stiffened grades of plastics materials provides numerous examples of what, in retrospect and with the benefit of current understanding, were often misguided, inefficient and expensive projects. The causes included the use of inappropriate tests, over-reliance on one particular test method, failure to identify critical parameters and inadequate rationales. There was also a collective unwillingness to concede that modification of a polymeric material with the objective of advantageously changing a particular property is generally likely to produce an undesirable change in another property. However, in much the same way as viscoelasticity theory influenced the way that modulus is evaluated, fracture mechanics has now influenced the way that strength and ductility are evaluated and the scope for erroneous trains of inference has been reduced, see Supplement S8.6.2 for instance. Even so, fracture mechanics theory is one aspect of continuum mechanics, and one should not expect it to be fully valid for filled plastics or polymer composites (which are, of course, filled plastics). That proviso is particularly pertinent for high volume fraction continuous fibre composites in which the large interface area and the likely presence of sharp internal cracks at debonded interfaces combine to frustrate a continuum theory. Otherwise, all that is now required is that experimenters should choose judiciously from the suite of test methods available to them.

S8.11 Non-destructive testing of composites in the aircraft industry

Non-destructive testing (NDT) of the composites used in aircraft structures is an essential constituent of the inspection procedures. Such materials/structures may be damaged or otherwise defective and yet show no outward or obvious signs of it.

Causes of in-service damage include:

1 water-cracking due to the freezing of water on the inside surfaces of air-craft structures, in particular the pressurised cabin;
2 damage from service vehicles. Most service vehicles are fitted with 'soft' pads, hence if they strike aircraft there is rarely any visible damage, though severe damage can occur on the non-visual (tension) side of composites.
3 Corrosion due to water ingress and, with carbon fibre composites, the electrolytic attack on metal structures to which they are attached;
4 Spread of inherent delaminations during service, e.g. minor manufac-turing faults may enlarge during service.

NDT techniques in common use in the industry are:

1 *'Low kV' X-rays.* Microfocus low kV (7–13 kV) X-rays is a through-transmission technique. It can detect fibre bundle breaks, resin-rich areas, voids, foreign objects. It cannot detect delaminations, dry areas, or identify the position of faults through the thickness of the com-posite. It has a 'resolution' of about 4% of component thickness.
2 *Ultrasonics.* This can be used in either a transmission or reflection mode. In either case, only faults in a plane at 90° to the direction of propaga-tion of the ultrasound can be detected, thus making the examination of 3D objects difficult. Delamination, voids, contamination can all be detected, and when used in the 'reflection' mode, the approximate posi-tion of the defect in the composite can be determined.
3 *Acoustic testing.* A contact probe emits some energy into a structure, and listens to the response of the structure in terms of amplitude and phase. It is said to detect any 'crack type' faults including delamination due to the acoustic loss such faults cause.

The techniques involve the propagation of compressional and shear waves at frequencies from 10 kHz to 50 MHz. Measurements are possible of velocity attenuation, scattering and reflection. Systems may be passive, requiring transducers or active as in acoustic emission under stress. The information obtainable falls into the following classes:

1   all elastic constants in the frequency range used;
2   all elastic constants of anisotropic or inhomogeneous materials. These include Poisson's ratios, torsional rigidities and volume compressibilities;
3   occurrence or development of porosity down to a fraction of the wavelength used;
4   onset of crazing or cracking;
5   state of bonding in two-phase systems;
6   mechanical failure.

The chief advantage from an operational point of view is the non-destructive nature and the potential rapidity and accuracy of the measurements, which may be displayed or recorded in a variety of ways.

The main difficulties which may be expected are:

1   high intrinsic attenuation, requiring the employment of suitable high power transducers and high gain amplifiers;
2   coupling problems between transducers and materials.

See reference 1 for a general summary for this particular industry. The techniques obviously have their use in strength testing in the laboratory where detection of incipient damage or otherwise undetectable internal damage is relevant to the investigation.

Reference

1 A. Mahoon. 'The role of non-destructive testing in the airworthiness certification of civil aircraft composite structures.' *Composites*, 19, 3, 229, 1988.

S8.13 Fracture toughness of fibre-reinforced composites

S8.13.1 Fracture toughness of discontinuous fibre reinforced injection-moulded plastics

The measurement of toughness of injection-moulded discontinuous fibre reinforced composites has been investigated using a linear elastic fracture mechanics approach. There are three particular problems that need to be addressed in order to proceed with a successful method.

1   The morphology of the sample will be characterized by the processing history. For injection moulding of fibre composites, the melt is delivered into a relatively cold mould under a shear stress field and then the mould is filled with the melt, experiencing an extensional stress field. This results in fibres being aligned in the direction of mould-fill at the surface of the mould, but aligned at right angles to the direction of mould-fill at the core of the mould. This fibre orientation, albeit an over simplification of reality, will have a significant influence on the initiation and growth of the crack.
2   These types of composite materials can strongly influence the initiation of crack growth. This presents some difficulties in identifying the force and energy absorbed by the specimen at crack initiation.
3   Composites are not homogenous materials. Therefore, the applicability of fracture mechanics theory is in question.

These issues have been addressed by ESIS Technical Committee 4 (Polymers and Composites),[1] who have developed a protocol for the determination of K_c and G_c. The measurements on which the protocol was based were conducted on a modified polyarylamide with 50% w/w glass fibre reinforcement. Plaque mouldings of thickness 2 mm and 5 mm were prepared and cracks were grown under carefully controlled conditions. Two specimen geometries, single edge notch bend and compact tension (see Chapter 9) were used and tests were conducted at speeds in the range 1–10 mm/min.

In addition to the fracture measurements, the ESIS TC4 project participants also measured some features on the fracture surface. In particular, it was important to determine the fraction of fracture surface that included an approximation to a 'smooth' fracture. Fracture toughness could then be objectively assessed from a plot of fracture toughness (e.g. K_c) against smooth fraction of fracture surface. This is shown in Fig. S8.13.1/1, where it can be seen that the value of fracture toughness varies widely.

Fracture toughness may be defined at some specific value of smooth fracture fraction (e.g. 0.5) and then this value can be used for materials comparisons. It can readily be seen that, without the clarifying influence of

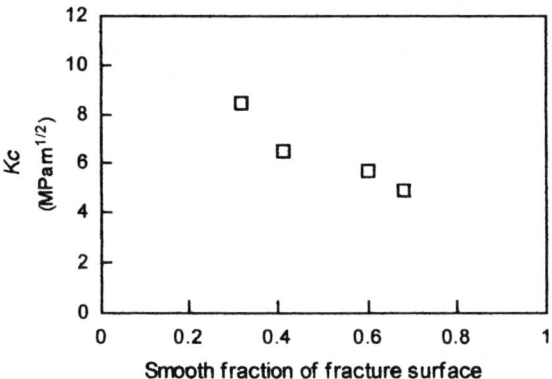

S8.13.1/1. Fracture toughness (K_c) plotted against the 'smooth' fracture fraction for 50% w/w glass fibre reinforced polyarylamide composite.

'smooth' fraction of fracture surface, the value quoted for toughness could take any value within a wide range. More important is the possibility of different criteria being used for different specimens and samples, which would lead to meaningless comparisons of toughness.

Reference

1 D. R. Moore, A. Pavan and J. G. Williams. *Fracture Mechanics Testing Methods for Polymers, Adhesives and Composites* Ch 1.3 'A protocol for the measurement of toughness for injection moulded fibre reinforced composites', D. R. Moore, Elsevier, 2001. ISBN 008 0436897.

S8.13.2 Delamination toughness of continuous fibre reinforced composites

Continuous fibre reinforced composite structures are prepared from prepregs of the resin and fibre and consolidated by heat and pressure to form an integral structure. The weakest fracture direction has been established to be between the plies and, if fracture occurs, then interlaminar or delaminating cracks will likely form part of the process. LEFM has been used to measure Mode I interlaminar or delamination fracture toughness. Much of the development has been conducted by ESIS TC4 and ASTM groups.[1,2]

An initial analysis of a double cantilever beam specimen geometry by Williams[3] provided a means of determining delamination toughness. An interlaminar crack was initiated from a PTFE foil located between the central plies of a unidirectional continuous fibre composite. LEFM concepts were combined with beam theory in order to provide an expression for the delamination toughness. The terms F, N and Δ relate to particular corrections associated with the test and are necessary for obtaining an accurate value for toughness.

$$G_C = \left(\frac{F}{N}\right)\frac{3P\delta}{2b(a+\Delta)} \qquad\qquad [\text{S8.13.2.1}]$$

where:

> F is a large displacement correction
> N is a load block correction
> Δ is a crack-length (a) correction
> b is the beam width
> P and δ relate to force and displacement

The term N relates to the stiffening of the beam due to the metal blocks that are attached in order to mount the beam onto a universal testing apparatus. A universal testing machine operating a ramp deformation function does not open both arms of a double cantilever beam specimen; instead, one is opened whilst the other remains stationary. This necessitates a crack rotation correction (Δ). Finally, a correction is sometimes necessary when a low stiffness beam exhibits large displacements of the beam (F).

Determination of fracture toughness generated from Eq. S8.13.2.1 provides values at initiation of cracking from the inserted PTFE starter crack and also crack propagation toughness values from an R-curve of G_C vs. crack length.

References

1 P. Davies, B. R. K. Blackman and A. J. Brunner. 'Towards standard fracture and fatigue test methods for composite materials'. ESIS Conference, Sheffield, 1998.
2 D. R. Moore, A. Pavan and J. G. Williams. *Fracture Mechanics Testing Methods for Polymers, Adhesives and Composites*, Ch 4.1 'Mode I delamination' B. Blackman, A. J. Brunner and P. Davies, Elsevier, 2000. ISBN 008 0436897.
3 S. Hashemi, A. J. Kinloch and J. G. Williams, 'The analysis of interlaminar fracture in uniaxial fibre–polymer composites'. *Proceedings Royal Society*, A427, 173, 1990.

S8.14 Databases on the strength of thermoplastics

A database on the strength of thermoplastics should reflect the potentially detrimental effects of the inherent viscoelasticity of the materials, the likely anisotropy of melt-processed end-products, and the presence of stress concentrators and aggressive environments. It is unlikely that an ideally comprehensive range of data currently exists for any plastics material, the cost of generating it would be deemed excessive and the wealth of detail might even confuse prospective users. Thus, it can be argued that a minimum contents database should be defined and also that a practical, user-friendly form of data presentation should be developed.

There have been some discussions between interested parties on various aspects of these matters but no definitive action has followed. However, it is likely that agreement could probably be reached about a first-stage development for tensile strength data along the following lines, using test pieces of standardized shape and dimensions except where otherwise indicated:

1   'Tensile strength' should be qualified as being either low strain brittle strength, yield stress or ultimate tensile strength, and clearly designated as such.
2   Strength measured at a standard temperature and straining rate by ramp excitation (i.e. in a conventional tensile test).
3   The same property measured at various other, preferably standardized, temperatures.
4   The same property measured at the standard temperature on thicker mouldings.
5   The same property measured on test pieces with an unfavourable flow geometry (preferably the most unfavourable one if such a lower bound can be identified).
6   The same property measured on double-edge-notched test pieces (notch tip radius 0.25 mm, notch depth 2.5 mm).

The reason for 4 is that the direction of molecular orientation varies through the thickness of an injection moulding. It is generally favourable to high axial strength in the skin layers and unfavourable in the core region of the end-gated injection moulded tensile bar, which is the commonly used and standardized tensile test piece. The values hitherto quoted in typical data sheets have generally been near the upper limit of attainable values because the thickness of the standard test bars has been such that skin layer orientation has predominated over core region orientation. The reason for 5 is similarly the need for realistic, service-pertinent data.

The wider aspects of tensile strength are not addressed in that tentative specification. Thus there are no references to creep rupture strength and fatigue strength, to the detrimental effect of certain environments or to fracture toughness. However, those omissions are largely tactical in view of the currently perceived enormity of the total task for strength data and the development history and current status of strength testing. This differs in significant ways from that of modulus testing, thereby instigating a different approach to the development of strength databases. In due course, tensile creep rupture data, for instance, could be included as derated 'tensile strengths' in the same way that creep modulus data can be presented as decimal fractions of a short term modulus (Chapter 6, Supplement S6.14).

Analogous issues arise for thermosetting resins, particularly those containing fibrous reinforcement and, for continuous-fibre composites, the lamellae stacking sequence can induce variations in strength similar to those caused by the flow geometry of thermoplastics. The data presentation issues for composites are similar to those for thermoplastics but the influential factors that have to be quantified are different. For instance, testing rate and temperature are much less important than the spatial arrangement of the fibres and lamellae. Additionally, the downstream data requirements of the designers, manufacturers and users of composite components and structures tend to be firmly aligned with well-defined property expectations for specific end-products whereas the data requirements for thermoplastics tend to be general and multipurpose, which is reflected in different database architectures and contents. Also, in the case of thermoplastics, there are protracted debates over the international standardization of presentation formats.

Strength and ductility from constant deformation rate tests

9.1

The design of conventional testing machines provides various constant rates of actuator movement which approximate to constant rates of straining. This would certainly be the case for many common test configurations when the strains are small. Historic equipment design employed analogue devices in order to control the actuator. However, in modern machine design the actuator is controlled with digital electronics, which makes complex input of stress or strain control readily accessible, including constant rates of increase of the applied force and constant rates of stressing if the deformation is uniform. Overall, the rationale is essentially the same as that which applies to constant rate of straining tests. Despite the much wider choice of input excitation in modern equipment, it is highly likely that the input format will continue to follow that devised in the past, since it is the design of the experiments (rather than the equipment) which dominates the test procedures. Thus modulus measurements at various constant rates of straining or stressing are feasible propositions (Chapter 4) but strength measurements usually entail larger deformations than modulus measurements and consequently depart more seriously from that ideal. They are often referred to as 'short-term' strengths, in contrast to the 'long-term' strengths, which relate to fracture or rupture after prolonged periods under stress.

The usual objective of a test at a constant rate of deformation is a single property datum, e.g. a modulus value or a strength, but the response curve, force vs. deformation, carries much more information than that. Constant deformation rate tests (and variants thereof such as that described in reference 1) have contributed outstandingly to the collective understanding of the mechanical behaviour of plastics. This is despite what was often crude experimentation using the earlier tensile test machines and the analytical inconveniences associated with ramp excitation (particularly when compared with step and sinusoidal excitations). Examples include areas as diverse as the micromechanics of necking[2] and the influence of molecular state of order on the tensile properties of films.[3] They continue to be one of the primary investigative tools for strength and ductility assessments though it can be argued that impact tests (Chapter 11) are superior for the ductility.

9.2

If the plastic is ductile in tension and breaks beyond the yield point, the datum commonly quoted nowadays is the yield stress rather than the ultimate strength. This is because the latter bears little relation to most service applications and also because the material often changes markedly from its original state as it is strained through and beyond its yield point. In addition, thermoplastic materials often suffer partial disruption via the formation of crazes, cracks or voids. When the test material suffers such change, the behaviour cannot be represented adequately by the current mathematical models of plasticity.

There is much experimental evidence, from creep studies and from tensile tests themselves, that with increasing strain the deformation processes become progressively dominated by molecular mechanisms that are either irreversible or reversible but with very protracted recovery times. The overall character of the deformation processes become 'viscous' rather than 'elastic' and the specimen then either extends uniformly or yields via a necking mechanism approximately in conformation with plasticity theory. The response curve of a ductile material in tension can generally be divided into four distinct regions (see Fig. 9.2/1) and its shape in Region 4 depends on whether the material work-hardens, work-softens or succumbs to necking rupture, but see Supplement S9.3. In the simplest case the yield point is defined as the maximum in the response curve (force vs. time or force vs. deformation), but if there is no peak in the curve a marked change in the slope may be taken as the indicator.

Fibre composite materials are in a class apart in that any distinction between brittle and ductile states is arbitrary and artificial. Failure seldom occurs as a genuine brittle crack progressing unhampered through the entire body of the item. It usually occurs by debonding between the fibres and the matrix, which often produces a response curve with an apparent yield point, followed by brittle fracture of the load-bearing fibres, which may have the same effect on the later part of the response curve as necking rupture. Many thermosetting resins are brittle but the incorporation of fibres often confers a pseudoductility; in contrast, thermoplastics in their ductile state are often embrittled by the incorporation of fibres or at least caused to fail by induced necking rupture. Thus, in isolation, the

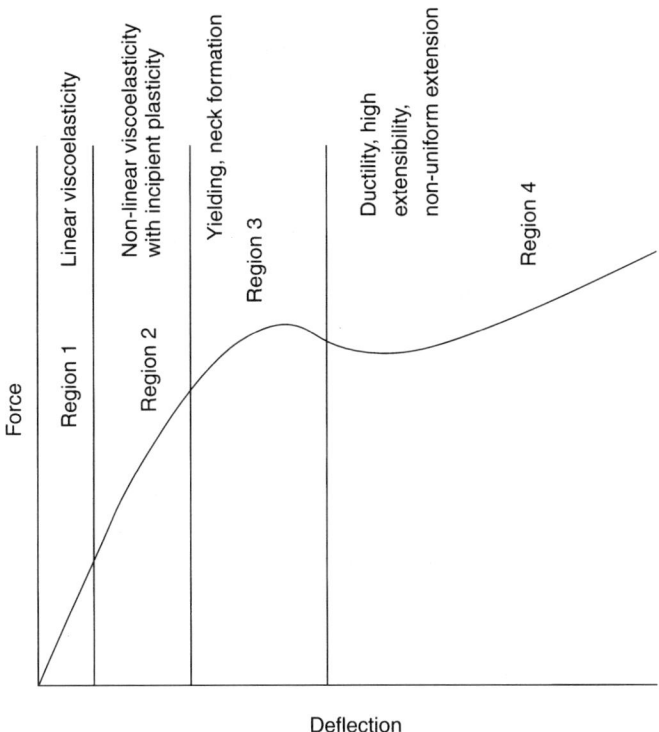

Force

Region 1 — Linear viscoelasticity

Region 2 — Non-linear viscoelasticity with incipient plasticity

Region 3 — Yielding, neck formation

Region 4 — Ductility, high extensibility, non-uniform extension

Deflection

Force – deflection   relationship to failure

9.2/1. Distinctive regions of the response of a ductile material to a ramp excitation. The shape beyond the yield point varies with polymer architecture and other factors.

apparently simple classification of failure behaviour of Fig. 9.2/1 could be misleading.

If the tensile strength, fracture or yielding, is plotted vs. temperature the curve generally has two distinct parts, see Fig. 9.2/2, corresponding to the two modes of failure. Ideally, at sufficiently low temperatures failures should be brittle and the strength should not vary with temperature. But, in general, the degree of brittleness and the strength do vary and the slope of the curve may be either positive or negative, depending on several inter-acting factors. At some point, the tensile strength begins to decrease more markedly with increasing temperature and the nature of the fracture changes to ductile.

At a different straining rate the curve is similar in shape but displaced, the transition to ductile fracture lying at a higher temperature for a higher straining rate. The transition from brittle to ductile behaviour is not usually

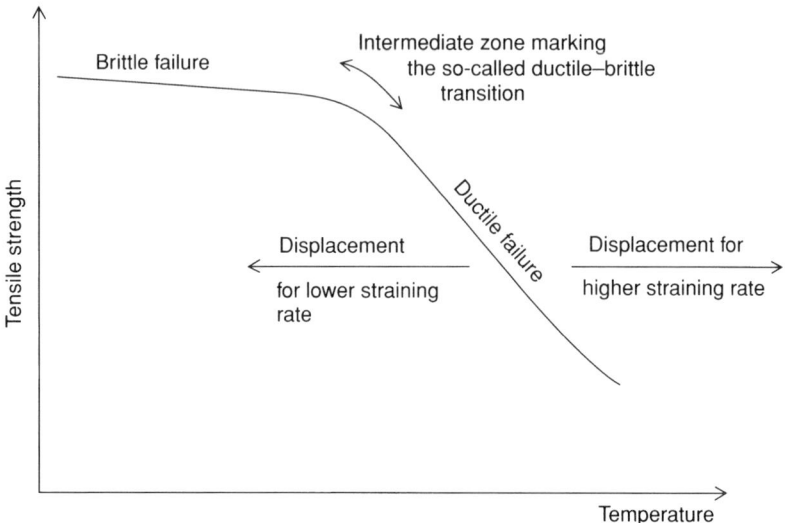

9.2/2. Tensile strength as a function of temperature. The position and shape of the curve varies with straining rate; the transition moves to higher temperatures as the straining rate increases.

abrupt, but a notional transition temperature can be identified by extrapolations from the regions in which the two states are clearly unambiguous. This process corresponds to the separate determination of K_c and σ_y in the derivation of ductility factor (Supplement S8.6.1). The position of the transition relative to room temperature governs our casual perception of a material as brittle or ductile.

Tensile yield stress is commonly measured and displayed as a function of temperature but not as a function of straining rate. It was found, for ramp loading times greater than 10 seconds, that a factor of 10 change in loading time (i.e. in deformation rate) has approximately the same effect on the tensile yield stress as a change of about 7 °C in temperature.[4] However, the quantitative shift varies with the polymer architecture. That approximation enables yield stress data over a range of straining rates, which are required for certain design calculations, to be estimated from the commonly available yield stress vs. temperature data.

9.3

The tensile strength in the brittle region is very sensitive to the 'quality' of the test piece or sample; surface flaws and internal defects will reduce strength. It can also be affected adversely by axial misalignment of the testing machine. Tensile yield stress is not comparably sensitive to imper-

Table 9.3/1 Tensile yield stress at 23 °C of common thermoplastics

Polymer type	Yield stress (MPa)
Nylon 66 (dry)	85
Polyethersulfone	85
Unplasticized PVC	55
Polypropylene	25
HD polyethylene	20
Polypropylene + 0.30 vol fraction of short glass fibres	85
Nylon 66 + 0.33 vol fraction of short glass fibres	180

fections and misalignments because local yielding redistributes stresses and thereby dissipates inadvertent stress concentrations and irregularities but it is affected by the molecular state of the sample, analogously to the modulus (see Supplement S9.3 for some practical details of tensile testing). The standard tabulations quote single values of strength measured on test pieces of a particular geometry under standardized conditions. A small selection of such tensile yield stresses is given in Table 9.3/1. The quoted values in the table refer to particular (unspecified) states of the sample. Some of the data, especially those for the samples containing short glass fibres, may therefore be unrealistic in that the alignments of the fibres or the polymer molecules may confer higher values than those likely to be manifest in most moulded articles.

The yield stresses of unreinforced thermoplastics in their glassy state seems to be limited to about 85 MPa or less. That is consistent with a short-term modulus of 3 GPa, a yield strain of about 0.05 and some viscoelastic non-linearity. The short-term modulus value of 3 GPa is typical of the so-called glassy state in which many plastics are nevertheless tough. The value is generally higher at lower temperatures, but the material then tends to be brittle. Brittle strengths measured in tension are unreliable, for the reasons given above.

Table 9.3/2 shows the strength-derating effects of certain flow geometries and stress concentrators for three thermoplastics containing short glass fibres. Those particular data relate to one batch of each compound moulded into bars, plaques and discs. The tensile strength datum normally quoted by the primary producers of such compounds is for the 0° bar. The tensile strengths of closely similar compounds from various producers differ only slightly from one another and the interbatch variability for one compound

Table 9.3/2 Tensile strengths and derating factors at 20°C

Specimen description	Nylon 66 + 0.33 weight fraction of glass fibres (dry)		Polypropylene + 0.25 weight fraction of glass fibres		PTMT + 0.20 weight fraction of glass fibres	
	Tensile strength (MPa)	Derating factor	Tensile strength (MPa)	Derating factor	Tensile strength (MPa)	Derating factor
0° bar[a]	187[b]	1	80	1	130	1
0° plaque	169	0.90	—	—	—	—
90° plaque[a]	95	0.50	~30	~0.34	—	—
0° disc	150	0.80	58	0.72	100	0.77
90° disc	112	0.60	79	0.98	76	0.58
90° disc with blunt notch (tip radius 2 mm)	107	0.56				
90° disc with sharp notch (tip radius 250 μm)	84	0.44	55	0.68	30	0.23
0° disc with sharp notch (tip radius 250 μm)	—	—	43	0.54	—	—

[a] 0° and 90° denotes the alignment of the test axis with respect to the main flow direction.
[b] The strength in flexure was 197 MPa.

is also insignificant relative to the large differences attributable to the flow geometries.

The elongation to final rupture in a ramp excitation tensile test is less meaningful than the strength, but it can be used as an index of ductility and/or as an indicator of freedom from flaws, particularly for films. It has no quantitative meaning whatsoever if the gauge region contains a neck or other discontinuity.

9.4

The interest in the strength of plastics under deformation modes other than tension is often motivated by an alternative route to estimates of the tensile strength of brittle materials. Flexure, uniaxial compression and diametral compression of a solid cylinder are all useful for that purpose, see Supplement S9.4. Flexure reduces the deleterious effects of machine misalignment and imperfections on the surface of the test piece and has the additional practical merit of being experimentally simple. Uniaxial compression often induces ductile failure in test pieces that would be brittle in tension and is the first choice method for obtaining a value of yield strength for thermosetting plastics, (see Supplement S9.4.2).

There can be some problems with the uniaxial compression of cylinders, first, because the actual stress field usually deviates from the ideal and second because specimens of all but the most ductile materials develop surface cracks which then propagate into the body of the specimen as additional surface cracks develop.[5]

Diametral compression is a stable configuration and is not very sensitive to specimen/machine alignment. Plane strain compression is another useful alternative to tension for ductile materials. The plane strain compression test avoids the plastic instability of the uniaxial compression test. A pair of flat parallel dies are forced into the opposite faces of a strip of the material; from the force and the thickness reduction the stress–strain relationship for large strains can be derived. Elastic forces and frictional resistance at the die/test piece interfaces produce errors in the results but those may be allowed for by extrapolation of the results for dies of different breadths.

Shear strengths are interesting from both the theoretical and practical standpoints, but there are many experimental difficulties if the plastic under investigation is in its glassy state or if the strains are large. Torsion is the simplest configuration and is satisfactory for certain composite structures. However, results for thermoplastics do not translate easily into shear data unless the test piece is a thin-walled hollow cylinder, in which case it is prone to torsional elastic instability and there is consequently a paucity of experimental results for the shear mode. Hassaini et al.[6] have developed a new shear test device for composites, which they call the 'cube test'. The test piece is a cube of side 18 mm. The grips and entire subassembly are so designed as to eliminate lateral forces and to minimize frictional resistance. Tests can be carried out under ramp and cyclic excitations.

9.5

The span/thickness ratio in a beam flexure configuration can be varied to change the ratio of shear stress to tensile stress. This is particularly useful for the measurement of interlaminar shear strength in certain classes of composite; if the span/depth ratio is small a shear failure criterion may be reached before a tensile or compressive failure criterion, see Supplement S9.5. The desired failure mode is not always attainable. The span to depth ratio cannot be reduced indefinitely because the rollers ultimately impinge. This can jam the beam in the loading device and other types of failure may intervene, for example, compression damage under the loading roller or buckling of fibres out of the compression face of the test piece.

The shear strength is derived from the equation:

$$\tau_{max} = \frac{3P}{4bh}$$ [9.5.1]

where P = force at failure
         b = width
         h = thickness

9.6

Tensile strength data for rubbers and elastomers are not really comparable with those in Table 9.3/1 because such materials usually extend to several times their original length before they break. The quoted values are not usually adjusted for the reduction in the load-bearing cross-sections, which in any case can rarely be measured accurately. Typically, tensile strength lies between 15 and 30 MPa and elongation to break lies in the region 300–600%.

Low molecular weight thermoplastics, others with a low yield stress and rubbers draw uniformly rather than through a neck. For some, the force–deflection curve turns upward at some point as new resisting mechanisms come into play. For example, as the extension increases, the molecules are progressively untangled to the point where further extension acts against the inherent stiffness of the molecules rather than against the friction between neighbours in the assembly. In some cases, particularly in certain rubbers, the increasing degree of molecular alignment permits the development of crystals which serve as crosslinks to reinforce the network.

9.7

The dominant feature of the tensile stress–strain behaviour of a ductile plastic is the degree of stability or otherwise of the neck. If the material cannot withstand the applied force as the cross-section decreases in the neck region the specimen will break there, the so-called necking rupture. For the neck to be stable the material has to become stronger, or to 'strain harden', as it orients in the neck to compensate for the decreasing section. The more effective this process, the greater the chance that the necked material will endure high extension before it breaks. Some light is thrown upon this issue by a construction allegedly due to Considère as discussed in Supplement S9.7. In fracture situations in service, necking rupture entails more energy absorption than brittle fracture. In addition, the energy density may be much higher in a necking region than that associated with a brittle crack where the total energy involved is usually quite small because the ductile failure zone is localized. The absorbed energy associated with necking rupture is far less than it would be if the material were to cold-draw and thereby involve a much larger volume of material in the energy-absorbing process.

9.8

Constant deformation rates are often employed in the measurement of fracture toughness. Common test configurations are:

- the single edge notched flexed beam (SEN)
- the compact tension specimen (CT)
- the double cantilever beam (DCB)
- the end-notched flexed beam (ENF)
- the edge-delamination tensile bar (EDT)
- the cracked lap shear specimen (CLS)

They serve different, though sometimes overlapping purposes. Some, for example the end-notched flexed beam, are uniquely suitable for tests on continuous fibre composites; others, for example the compact tension specimen and the single edge notched beam, are suitable for all plastics materials. Nevertheless, apart from a few notable exceptions, such as the SEN beam, the configurations entail somewhat troublesome machining of the test pieces or special moulding procedures and considerable testing expertise. Additionally, they all pose the common problem of reproducibility of the sharpness of the starter crack. Those severely practical difficulties tend to inhibit exhaustive experimentation which, in turn, has limited the pool of information on attainable accuracies and precisions, reliability, etc. particularly in relation to anisotropy in thermoplastics mouldings (fibre-reinforced plastics components have received special dedicated attention). However, developments continue. A modified compact tension configuration utilizing an edge-gated injection-moulded disc has been useful for anisotropic thermoplastics samples because the loading points and the notch can easily be variously positioned with respect to the main flow axis.[7] Discs are used for many test purposes in the thermoplastics industry and therefore wide adoption of the modified compact tension configuration would link fracture toughness with other mechanical properties unambiguously. Meanwhile, for rapid assessments, flexure has the practical advantage over tension in that there is no clamping. It is also the more severe configuration because the neutral axis in bending is a restraint to plastic deformation. See Supplement S9.8 for some practical details and also results, from tests on beams cut from a picture-frame cavity, from which realistic, service-pertinent data may be obtained.

9.9

Even plastics that are widely regarded as being outstandingly tough or that are remote from their tough–brittle transition as indicated on their conventional strength vs. temperature relationship (see Fig. 9.2/2), can be embrittled by an appropriate combination of test piece geometry and test conditions. The sharp notch or crack employed in fracture toughness testing is one example of the geometry effect. Depending on the dimensional details, the stress field in the region of the crack tip may be triaxial, which favours brittleness because the molecular responses to an excitation are retarded under triaxial conditions. Increased straining rate is similarly effective and, additionally, the geometry can affect the straining rate (e.g. at a crack tip it is much greater than the overall deformation rate would imply for an uncracked specimen).

9.10

A practical difficulty at high straining rates is noise generated in the system, which can distort or obscure the true response curve. The primary data on peak force, for example, may be severely erroneous. Electrical filtering can remove the extraneous vibrations but may simultaneously suppress some genuine features in the response. On the other hand, mechanical damping will change the contact stiffness between actuator and specimen and merely increase the time to fracture.

Extraneous vibrations can be particularly troublesome if the material is in its brittle state. However, they are generally insignificant if the test temperature and the natural frequency of the system coincide with a region of high internal losses (high $\tan \delta$) in the material, see Chapter 5, or if the material is ductile. The ductile state also entails high internal losses that arise from large-scale molecular movements rather than the segmental movements responsible for phase lag in low amplitude sinusoidal excitations.

Inevitably, distorted response curves are especially likely in impact tests, see Chapter 11.

9.11

Tensile yield stress has a unique and versatile role as a characterizing quantity, in that

1 it is a standard property that is widely measured;
2 it is a bridge between the phenomena of modulus and ductility;
3 some fraction of yield stress could be substituted for isometric stress in a design calculation;

4 some fraction of it could be substituted for creep rupture stress in design calcuations to avoid long-term failure (see Chapter 10);

5 σ_f/σ_y could be used as a derating factor for creep rupture.

A note on tensile yield strength as a datum for engineering design is included as Supplement S9.11.

9.12

Apart from the use of various conventional test piece geometries referred to earlier, it can be argued that new shapes and test configurations should be developed, to enable strength anisotropy and the potentially detrimental effects of flow irregularities to be studied easily. As discussed elsewhere, plate-like test pieces are the simplest new shape and plates with various flow geometries have an established usefulness for the assessment of mechanical properties such as modulus and impact resistance. Their adoption for some strength testing brings the same advantage of greater downstream relevance and the additional advantage of integration with the other mechanical property data. Some plate test results are given in Supplement S9.12.

References

1 H. Doubravszky. 'How to get more information from a standard stress–strain curve of plastics.' *Plastics, Rubber and Composites: Processing and Applications* 15, 3, 177, 1991.

2 A. Cross and R. N. Haward. 'Post-yield phenomena in tensile tests on poly(vinyl chloride).' *Polymer*, 20, March, 288, 1979.

3 R. J. Rolando, W. L. Kruger and H. W. Morris. 'The influence of strain rate on tensile properties of polypropylene films.' *Plastics and Rubber: Processing and Applications* 11, 3, 135, 1989.

4 P. I. Vincent. Private Communication.

5 R. Papirno, J. F. Mescall and A. M. Hansen. 'Fracture in axial compression of cylinders' in *Compression Testing of Homogeneous Materials and Composites*. R. Chait and R. Papirno, Ed. ASTM STP 808, 1983, p. 40.

6 D. Hassaini, E. Vittecoq and G. Degallaix. 'Study of monotonic shearing behaviour of unidirectional glass-epoxy composite using new testing device.' *Plastics, Rubber and Composites: Processing and Applications* 27, 5, 227, 1998.

7 R. D. Whitehead, D. R. Moore, D. C. Leach, K. E. Puttick and J. G. Rider. 'Fracture testing of polyethersulphone injection mouldings.' *Plastics and Rubber: Processing and Applications* 8, 2, 115, 1987.

S9.3   The direct measurement of tensile strength

Introduction

The conventional tensile strength testing of plastics follows a procedure that was initially developed for tests on metals and often conforms to one or other of several Standard Methods or to a Code of Practice. The particular type of excitation traditionally used in tensile testing was originally chosen for its practical convenience and mechanical simplicity. It approximates loosely to a ramp function of strain (or stress in some tests) vs. time, deviation from the intended function arising because a perfect ramp would entail an abrupt transition from rest to a finite and constant rate of stressing or straining. That departure from a true ramp at the onset may affect the accuracy of modulus data, but is usually unimportant in strength tests because constant rate of deformation has been attained before the specimen breaks.

Most of the stipulations set out in the standardized practices embody the collective wisdom of earlier tensile test practitioners. In common with most mechanical test procedures, they fall into a number of distinct groupings, including at least the specimen–machine system, the precision of the data and the physical interpretation of the data.

The test piece–machine system is obviously the primary consideration because unless it functions properly no worthwhile data can be generated. The other stages are supplementary but nevertheless essential in that they enable the outcome of the specimen–machine interaction to be translated first into test data and then into mechanical properties data for the sample under investigation. Additionally, all stages are strongly affected by the inherent nature of the samples and specimens, i.e. the viscoelasticity of almost all classes of plastic, the heterogeneity of most of them and the tendency for test pieces and other end-products to be anisotropic and to contain textural irregularities, etc. The mode of failure may variously be by brittle fracture, necking rupture, gross yielding or after extensive cold-drawing. This raises questions as to how strength should be defined, described and measured. The variety of strength and failure characteristics also necessitates a high level of versatility in both the test machines and the evaluation programmes.

The practicalities of the tensile testing of anisotropic specimens of viscoelastic materials

Viscoelasticity and anisotropy compromise the essentially simple functional operation of the test machine and the subsequent interpretation of the raw data. Starting with the specimen–machine interaction, the force is transmitted to the specimen mainly via shear stresses at or near the grips. However, many materials have a deliberate predominantly axial molecular orientation or fibre alignment, which confers a relatively high tensile strength and a relatively low shear strength along the longitudinal direction. This results in possible shear failure near the grips before tensile failure in the gauge region. Modified grips, reinforcing plates attached to the ends of specimens and changed specimen profiles can all reduce the risk of such malfunction. However, such steps may be detrimental in other respects, for instance, if the predominant orientation lies at some angle to the tensile axis then the specimen distorts into a sigmoid, the exact form of which depends on whether the clamped ends are free to rotate or not, and in either case the observed force and extension do not necessarily convert into correct values of tensile strength. In general, those clamping effects are only severe in test pieces from thermoplastics foils or films and from continuous fibre reinforced composites, but even when there is no gross malfunction mildly erroneous test data are likely to be generated if appropriate corrective action is not taken.

Viscoelastic test pieces also creep in the grips and slip giving errors in failure strain as estimated from actuator movement, but see later, and may distort under the attachment forces of an extensometer if one should be used. In general, viscoelasticity combines with mechanical inertias and finite response times of the sensors to distort the observed force–deformation relationship. This can cause a reduction of the initial slope, obscure the origin and also change features of the response curve that might relate to structural changes in the deforming test piece, thereby detracting from the usefulness of the test and introducing possibilities for errors in the measurements. Above all, however, the response reflects both the increasing strain and the passing of time as the test progresses and therefore the observable response is not as immediately informative as the response curves to step excitations, say a creep curve or a stress relaxation curve. A force–deformation or stress–strain record contains no direct indication of load-bearing capability under loads sustained over any period of time greater than the duration of that particular tensile test, though there may be indicators that a skilled observer may discern and interpret. Tensile testing practice accommodates the two deficiencies pragmatically by regarding deformation rate, rather than time, as a critical variable. Therefore, comprehensive evaluation entails the use of straining rates ranging over several decades, though the range is limited at the slow end by considerations of cost effectiveness and at the fast end by practical features of the test machine.

In summary, viscoelasticity and anisotropy can distort the specimen–machine interaction, cause extraneous deformations in specimens under test, induce failure at or near the grips and obscure the true response to ramp excitation to the point where generated data may be misleading. Evaluation programmes may have to be expanded to serve their intended purpose.

The measurement of force and deformation

The practical experience accumulated in 200 years of conventional tensile testing of many classes of material and 100 years of tensile creep testing have led to a design specification that includes the following, even for samples and specimens that are neither viscoelastic nor anisotropic:

1   adequate power in the testing machine, to ensure that the stiffest specimens can be extended at the designated rates;

2 alignment of the line of action with the axis of symmetry of the specimen, to minimize the variation of stress across the specimen cross-section;

3 secure and balanced clamping of the specimen to ensure that it neither slips in the grips nor suffers extraneous forces;

4 high quality specimens of the correct size and profile for the intended purpose and with an adequate surface finish.

Those four design features are interconnected to some degree.

The provision of adequate power poses no direct problem. However, there may be secondary difficulties in that a powerful machine is likely to be massive and to have inertias and frictions in the actuator and the linkages that are troublesome when the active forces are small, e.g. in the early stages of a test on a plastics material. Ideally, the choice of machine should be such that the machine is matched in its mechanical design to the properties of the test pieces. The latter should never dominate the machine. This is because the extracted signal would reflect a complex combination of machine and specimen characteristics; on the other hand, if the machine dominates excessively, it may impose inadvertent and undesirable constraints on the response of the test piece.

Accurate alignment of the test piece in the machine is necessary for the best performance, but is not easily achieved because the machine, the test piece and the linking of the one to the other are all prone to asymmetries that can cause misalignment. The design choices lie between sufficient degrees of freedom to allow automatic realignment as the test piece extends, at one extreme, to total constraint at the other. The former method relies on the test piece being sufficiently stiff to be essentially unaffected by the adjustment forces, which is unlikely to be the case for a plastics material. Similarly, however, the friction almost inevitable in a fully constrained system may constitute a large error on the measured force.

Even if the specimen is satisfactorily symmetrical about its longitudinal axis, which requires special provisions at the production stage, it may be clamped unsymmetrically unless special precautions are taken to locate it properly in the grips. A hole in each specimen end and corresponding pins in the grips is the simplest solution, which has been very satisfactory for tensile creep tests, provided that anisotropy does not simultaneously confer a high axial strength and a low shear strength along the same direction; the holes also facilitate any machining operation by defining the axis of symmetry. Ideally, the force should be transmitted to the test piece solely via the pins; the less ideal conventional clamping, ostensibly acting across the entire width of the specimen, is commonly used.

To some degree there is a conflict of objectives in the design and operation of the grips. Secure clamping is necessary so that the test piece does not slip relative to the grips during a test, but it simultaneously prevents self-aligning movement and thereby preserves any initial misalignment. On balance, total constraint is the preferred option, in which case hydraulic grips are probably the most satisfactory because they exert a uniform pressure over the whole face which remains constant as the test piece extends and thins correspondingly. Simple mechanical grips may have to be over-tightened initially and the test piece may consequently distort badly there or premature fracture may be induced there. Such possibilities can be reduced or eliminated by the use of reinforcing tabs on the ends of the test pieces, but this is a tedious measure that is not widely used for tests on plastics but is virtually obligatory for tests on continuous fibre reinforced composites. Furthermore, the issues of alignment and malfunction at the grips are less important for large test pieces machined from thick stock, but that is rarely an option with plastics and composites.

The experimenter has no option but to measure the force by some remote sensor and deformation either at a surface or over a finite gauge region. However, the physical characteristics of the specimen and the material should be expressed in terms of stress and strain, often at a local site of failure initiation. Thus, the translation of force into stress and deformation into strain is a source of errors and approximations, so much so, that the transformed results may bear little relation to the fundamental properties. However, that does not render the tests useless because much can be inferred qualitatively from the post-yield shape of a force–deformation curve and even quantitatively from the elongation to break when correlations can be established between that quantity and sample quality (for a particular class of material tested in a specific way). Even so, Note No. 2 of ASTM D638 *Standard test method for tensile properties of plastics*, states cautiously, apropos of other factors but appropriately nevertheless, 'This test method is not intended to cover precise physical procedures. . . . Special additional tests should be used where more precise physical data are required.'

The force is always measured directly, and accurately provided the machine and the transducer are adequately stiff. The sensors should have sensitivities and response times that are commensurate with the intended purpose of the test. They should be able to discriminate at, say, 0.01 of full-scale display, and the response time should be such that the fine structure of the response is detected even though extraneous vibrations in the machine may then distort the observed responses.

Up to the yield point the deformation may be measured directly via an extensometer attached to the specimen, strain gauges bonded to it or an independent optical device operating without physical contact. Only the remote action optical devices are practicable beyond the yield point and, for most tests in the ductile state, extensometry as such is dispensed with, the deformation then being measured indirectly as actuator movement (with possible corrections for extraneous effects due to clamping and the specimen profile).

The overall precision of measurements of strength is rather less than that attainable for modulus measurements. If the failure is brittle, the calculated strength can be based on the initial cross-sectional area, but that measured quantity may be neither precise nor accurate because of non-axial loading, defects in the specimen or variable anisotropy. Coefficients of variation of 0.10 for the interspecimen variability are commonplace and a measured value may be a substantial underestimate of the true strength. If the failure is ductile, non-axial loading and specimen defects are less important, but the estimate of cross-sectional area is likely to be erroneous and if the deformation is also inhomogeneous, as is common, the calculation of failure stress is further confounded. The nominal yield stress calculated on the basis of the initial cross-section is likely to be precise, with a coefficient of variation of about 0.03. It is not highly accurate because of the complexity of the associated phenomena and uncertainties as to what point or feature on the response curve should be regarded as indicative of the onset of yielding. Whatever point is taken, the associated nominal yield stress should not be regarded as unambiguously definitive because there is a zone in which the material is neither wholly viscoelastic nor wholly plastic and, additionally, the material in the neck is not necessarily a continuum. Another potentially important factor is the temperature in the zone of yielding which, depending on the deformation rate, will tend to rise above the ambient temperature of the test. It may reach a critical point at which the inherent yield stress has fallen to such a level that the neck cannot support the prevailing force. If the neck stabilizes satisfactorily, it will travel along the parallel-sided section of the test piece either at an approximately constant force, or with a progressively increasing force if the molecular assembly favours further orientation, until it reaches a flow irregularity or a discontinuity which promotes failure.

A nominal breaking stress beyond the yield point is little more than a normalized breaking force, is physically meaningless and is useful only as an indicator in comparisons with similar data for other samples.

The deformation or strain at failure beyond the yield point is similarly a dubious quantity. It is usually inferred from the movement of the actuator because strain gauges and extensometers are not normally used in tests that are expected to progress to failure of the test piece, and remote sensors are not reliably accurate under typical test laboratory conditions. With brittle fracture, the error on the inferred deformation is usually large because extraneous deformations at, and near, the grips constitute a relatively large proportion of the overall movement of the actuator. That source of error is less influential when the test piece yields but the measured deformation does not then usually translate directly into strain. Even so, the commonly quoted extension to break is a useful datum because it relates loosely to the stability of the neck, the propensity of the specimen to subsequently work harden and the incidence of defects in the specimen. There are no quantitative rules to underpin judgements on those matters and the investigator has to assess new results against a background of whatever accrued data are deemed appropriate.

The same is true, to varying degrees, of most of the data relating to failure; they are accommodated within a framework of comprehension that enables useful information to be extracted despite uncertainties about the physical credentials of the experimental data. That framework of comprehension is based on the collective experience of many previous investigators, accumulations of data, established correlations between results from test samples and service performance, perceptions of quality and other knowledge. It follows that the reliability of such rationalizations depends heavily on the quality of the database.

The principal sources of error that are encountered in this phase of the testing operation, in combination, often lead to coefficients of variation of 0.10 or more. At that level, the imprecision is such that at least ten nominally identical specimens should be tested for the derivation of a property value (most standard specifications stipulate a minimum of five).

S9.4   The indirect measurement of tensile strength

S9.4.1 By flexure

The tensile strength of brittle plastics is often measured by three-point flexure. Ostensibly, this circumvents the premature failure that can arise in tensile tests for the reasons listed in Supplement S9.3. In addition, the configuration is easy to set up. However, the apparatus for strength measuremesnt is not as simple as that for modulus measurements. In the latter the radii of the loading bar and the supports can be small, because the applied forces are low, and consequently the span is well defined. However, strength tests entail higher forces and the radii have to be larger to avoid local damage to the beam and consequently the span is less well defined as the beam flexes. Various radii are permitted by the Standard Methods. The general consensus seems to be that the minimum radius should be 3 mm, the maximum radius of the support rollers should be 1.5 times the thickness of the beam and the maximum of the radius of the loading nose should be four times the thickness.

The linear elastic formula is:

$$\sigma_m = \frac{3PS}{2bh^2} \qquad\qquad [\text{S9.4.1.1}]$$

where σ_m = stress in the tensile skin layer at mid-span
         P = force at failure
         S = span
         b = width
         h = thickness

which is valid only if the mechanical properties in tension and compression are identical, if the span is large relative to the thickness (span/thickness ratio 20 or greater), if the central deflection is very small and if the total length of the beam is not much greater than the span.

Modified expressions can be used if that deflection is not small, viz. BS 2782 Part 10,

$$\sigma_m = \frac{3PS}{2bh^2}\left[1+4\left(\frac{\delta}{S}\right)^2\right] \qquad\qquad [\text{S9.4.1.2}]$$

where δ = central deflection

or ASTM D790M,

$$\sigma_m = \frac{3PS}{2bh^2}\left[1+6\left(\frac{\delta}{S}\right)^2-4\left(\frac{\delta h}{S^2}\right)\right] \qquad\qquad [\text{S9.4.1.3}]$$

A high span/thickness ratio ensures that failure is tensile if the deflection is small, but it invalidates the formula if the deflection is large. The reaction forces at the supports are then not parallel to the line of action of the applied force and surface tractions at points of contact and membrane stresses may arise.

Overall, the flexure test is not suitable for plastics that fail in a ductile manner nor is it an accurate strength test for brittle materials (in contrast to its precision, and possibly accuracy, for modulus measurements). On the other hand, when carried out with due care tensile strength data derived in flexure tend to be genuinely higher than those derived in tension. Factors contributing to that disparity are differences in the way cracks grow in the two deformation modes and the greater volume of material, and hence the greater number of incipient flaws, that is exposed to the maximum stress during a tensile test.

S9.4.2 By uniaxial compression

The yield strength of a brittle thermoplastic and almost all thermosetting plastics cannot be measured accurately in a tensile test and therefore it is necessary to conduct a compressive test to generate a yield characteristic. One particular procedure that has grown in value is unixial compression of block type specimens with a square prismatic cross-section (sides of length a and height l).

A relationship exists between applied stress (σ_a) in compression and true stress (σ)[1]:

$$\sigma_a = \sigma\left(1 + \frac{\mu a}{l}\right)$$

[S9.4.2.1]

where μ is the coefficient of friction between the specimen and the anvil of the test machine. Equation S6.5.1 in Chapter 6, which refers to axial compression of a cylindrical prism, is the same except for a different numerical factor.

If the test piece has a high slenderness ratio, buckling may occur before the test piece yields and if the ratio is small the test piece will not buckle but friction at its ends may inhibit lateral movement and the test piece may then 'barrel'. A grease at the interface should reduce the friction effect. The usual practice is to test several test pieces of different transverse dimensions but with no buckling. A plot of measured compressive yield strength vs. a/l extrapolated to $a/l = 0$ then enables the true compressive yield strength to be determined.

Tensile yield strength can then be derived from the compressive value by application of plasticity theory[2] in the following manner:

$$(\sigma_y)_{\text{Ten}} = \frac{(\sigma_y)_{\text{Comp}}}{1.3} \qquad\qquad [\text{S9.4.2.2}]$$

This relationship has proved to work for a wide range of materials. The factor 1.3 is empirical in origin although it has some basis in plasticity theory.[2]

References

1 M. Davies and D. R. Moore. 'Laboratory-scale screening of mechanical properties of resins and composites: relevance to composites for aerospace applications.' *Composites Science and Technology* 40, 131, 1991.
2 S. Gali, G. Dolev and O. Ishai. 'An effective stress/strain concept in the mechanical characterization of plastics.' *International Journal of Adhesion and Adhesives* Jan 1981, 135.

S9.5 Shear in short flexed beams

The short beam shear test configuration uses a small span-to-thickness ratio in the three-point bending configuration to bias the stress field so that shear failure rather than tensile or compressive failure occurs. The test configuration and the stress field for a homogeneous, isotropic beam are shown in Fig. S9.5/1. The through-thickness shear stress distribution for three-point loading is given by:

$$\tau_{zx} = \frac{P}{4I}\left(\frac{h^2}{4} - z^2\right)$$ [S9.5.1]

where τ_{zx} = shear stress
   P = force
   I = area moment of inertia
   h = thickness
   z = distance from neutral axis

The relationship is parabolic; the maximum value of τ_{zx}, which occurs in the mid-plane of the beam at $x = 0$, is $\dfrac{3P}{4bh}$ and the value is zero at the surfaces.

The maximum of the tensile stress and of the compressive stress occur on opposite surfaces of the beam and, if the two stresses are equal, is given by:

$$\sigma = \frac{3PS}{2bh^2}$$ [S9.5.2]

where S is the span and the mode of failure depends on which breaking stress is exceeded first as the beam is flexed. The equations given here are only approximations. They are strictly valid only for large span-to-thickness ratios and also in the short beam configuration the stress field is distorted under the loading nose and at the supports. The shear stress distribution is consequently not uniform and symmetric about the central plane and the maximum value is not located at $z = 0$.

The test is often used on continuous fibre reinforced composite laminates for the measurement of interlaminar shear strength. This is because the typical degree of anisotropy manifest by such systems is such as to favour shear deformations (Eq. S3.2.2.2 in Supplement S3.2.2). The actual behaviour of such structures deviates from the idealization represented by Eq. S9.5.2 to a greater degree than does the behaviour of homogeneous, isotropic beams. The through-thickness variation of shear stress is not smoothly parabolic as implied by Eq. S9.5.1. This is because of the structural discontinuity at the interface between each lamella and when the lamellae are stacked at various angles, which is commonly the case, torsion–shear coupling occurs, normal forces arise across the interfaces and

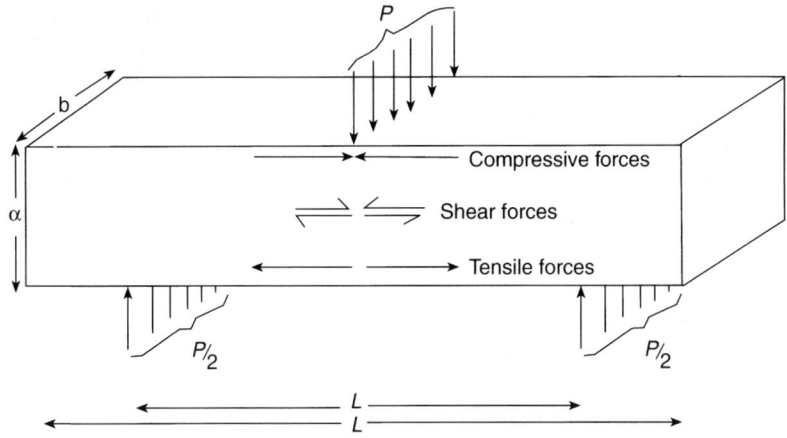

Shear stress distribution for a homogeneous beam

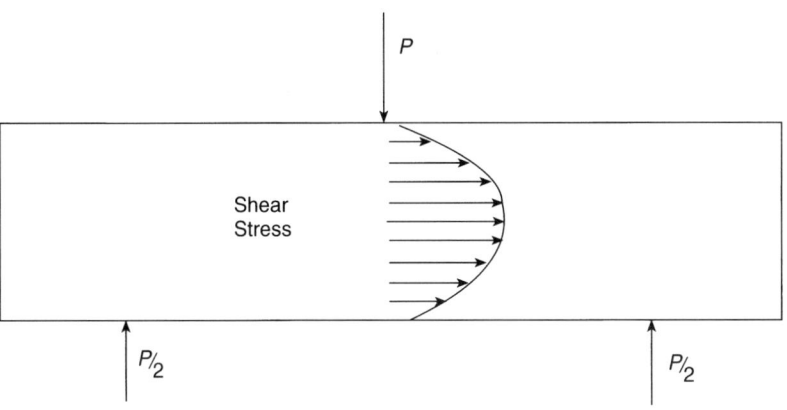

Shear stress distribution for an inhomogeneous laminated beam

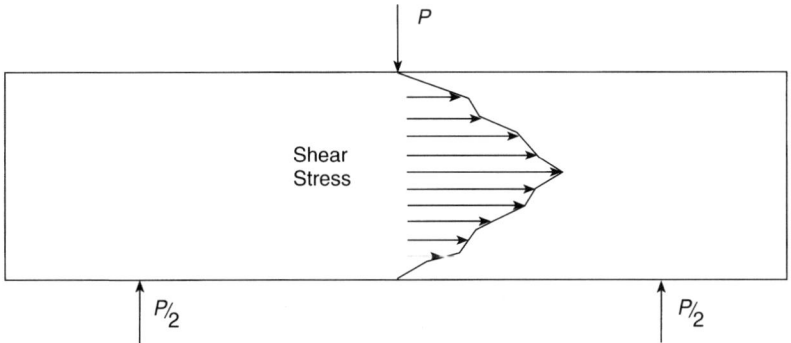

S9.5/1. Short-beam shear test configuration and the forces acting.

failure modes other than interlaminar shear can occur. Additionally, laminating imperfections at the interfaces, for instance voids and out-of-plane fibres, can complicate the outcome of a test, see reference 1 for instance. Some defects reduce the strength, but misaligned fibres can increase the measured value of apparent interlaminar shear strength even to values higher than the shear strength of the matrix. However, the tensile strength and the tensile modulus is correspondingly lowered in such instances, which exemplifies the informative usefulness of certain datum pairs that was referred to in Chapter 2.

Failure can initiate also at fibres buckling out of the compression face of the beam and an interfacial crack can distort the stress field to the point that cracks are induced along other interfaces. Thus, it is imperative that any broken test piece should be inspected and the strength datum rejected as an interlaminar shear strength if there is visual evidence of other failure modes.

The short beam shear configuration can be used also on plastic–plastic laminates. For example, it has been widely used on sandwich foam injection mouldings in order to determine the shear strength at a skin–core boundary.

Reference

1 S. Singh and E. Greenhalge. 'Micromechanisms of interlaminar fracture in carbon fibre reinforced plastics at multidirectional ply interfaces under static and cyclic loading.' *Plastics, Rubber and Composites: Processing and Applications* 27, 5, 220, 1998.

S9.7 Considère's Construction

The deformation of plastics near to, and beyond, the yield point may be considered as taking place at constant volume and therefore:

$$Al = A_0 l_0$$

and $l/l_0 = 1 + \varepsilon$

where A = cross-section
        l = length
        ε = strain
and subscript zero refers to the initial length and cross-sectional area.

The true stress σ is force/A and the notional (net) stress σ_n is force/A_0, and therefore,

$$\sigma = \sigma_n (1+\varepsilon)$$

and

$$\frac{d\sigma_n}{d\varepsilon} = \frac{1}{(1+\varepsilon)^2}\left[(1+\varepsilon)\frac{d\sigma}{d\varepsilon} - \sigma\right] \qquad \text{[S9.7.1]}$$

Therefore, a stress–strain curve plotted in the standard manner, with stress calculated on the basis of the initial cross-section, i.e. the notional stress, will have zero slope when:

$$\frac{d\sigma}{d\varepsilon} = \frac{\sigma}{1+\varepsilon} \qquad \text{[S9.7.2]}$$

At that point the slope of the true stress vs. strain curve is equal to the tangent to it drawn from a point with coordinates $(0, -1)$, see Fig. S9.7/1(a). One can distinguish three cases:

1   $\dfrac{d\sigma}{d\varepsilon} > \dfrac{\sigma}{1+\varepsilon}$, which corresponds to stable elongation;

2   $\dfrac{d\sigma}{d\varepsilon} = \dfrac{\sigma}{1+\varepsilon}$ on one occasion, which corresponds to the formation of a neck (a maximum on the conventional stress–strain curve) followed by thinning and failure;

3   $\dfrac{d\sigma}{d\varepsilon} = \dfrac{\sigma}{1+\varepsilon}$ on two occasions, which corresponds to formation and subsequent stabilization of the neck, high elongation to break and at least some work hardening.

(a)

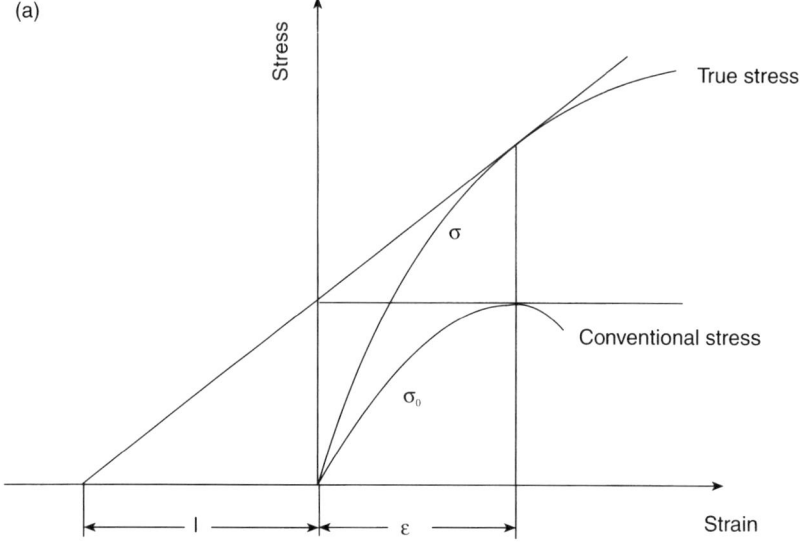

(b)

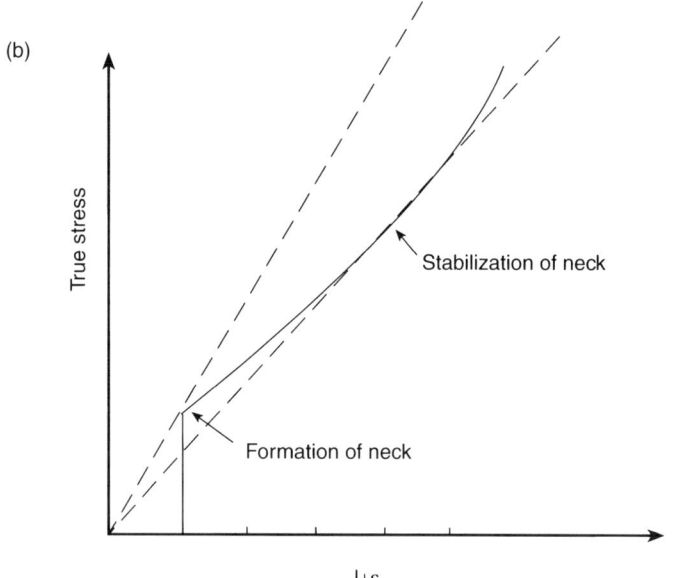

S9.7/1. Schematic tensile stress strain curves according to Considère's Construction.
(a) Considère's Construction.
(b) True stress vs. draw ratio for a material that necks and subsequently work hardens.

It follows that an effective presentation form for stress–strain data beyond the yield point is true stress vs. draw ratio, i.e. $(1 + \varepsilon)$, see Fig. S9.7/1(b). That form has been used for the presentation and analysis of tensile data on film samples, but the common method of presentation for other samples remains either force vs. elongation or notional stress vs. engineering strain.

S9.8 Constant straining rate fracture toughness testing

S9.8.1 Stress field intensity factors and ductility factors from slow flexure tests

For three-point flexure, the stress field intensity factor in an opening crack mode is given by:

$$K_{lc} = \frac{3}{2}\frac{PS}{bh^2} Ya^{1/2} \qquad\qquad [S9.8.1.1]$$

where P = force at fracture
         S = span
         b = specimen width
         h = specimen thickness
         a = notch depth

$$Y = 1.96 - 2.75\left(\frac{a}{h}\right) + 13.66\left(\frac{a^2}{h}\right) - 23.98\left(\frac{a^3}{h}\right) + 25.22\left(\frac{a^4}{h}\right)$$

If P is plotted against $\dfrac{2bh^2}{3SYa^{1/2}}$ for test pieces with various notch depths, the graph should be a straight line of slope K_{lc} if the fractures are brittle and linear elastic. When the critical force cannot be measured accurately (as may be the case if the straining rate is high, under impact conditions, for instance) an alternative analysis uses G_{lc}, the critical strain energy release rate. That is calculated from the expression:

$$G_{lc} = \frac{U}{bh\phi} \qquad\qquad [S9.8.1.2]$$

where U = impact energy required for fracture
         ϕ = a geometric factor[1]

In most applications for plastics, the subscript I is omitted since a crack opening mode is assumed.

The stress field intensity factor may then be calculated from the relationship

$$K = \frac{EG}{(1-\upsilon^2)} \qquad\qquad [S9.8.1.3]$$

In order that the values of K_{lc} or G_{lc} pertain to plane strain conditions, it is necessary that the specimen width, b, exceeds a minimum value (b_{min}) given by:

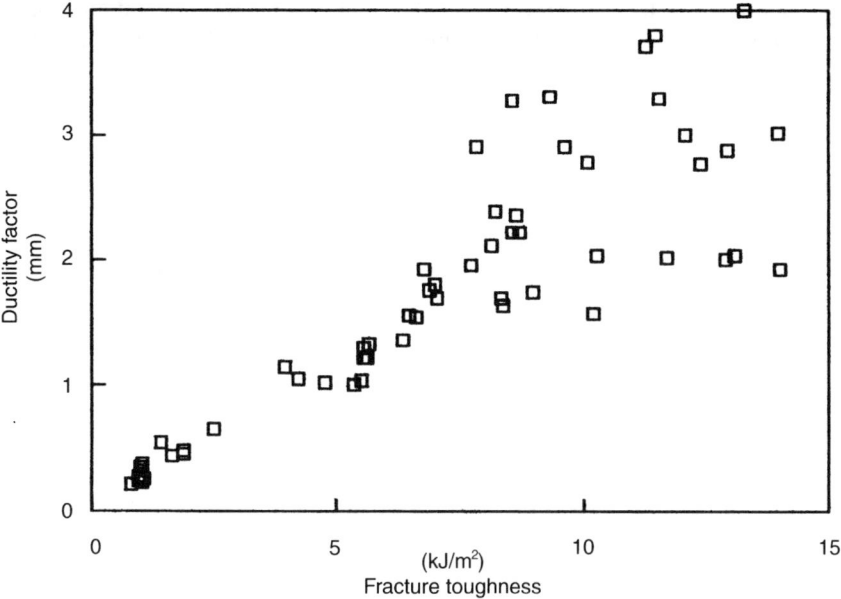

S9.8.1/1. Ductility factor vs. fracture toughness for some ABS compounds tested over a range of temperatures.

$$b_{min} = 2.5 \left(\frac{K_C}{\sigma_y} \right)^2 \qquad\qquad [S9.8.1.4]$$

and usually that $2b = h$

The importance of deriving plane strain geometry independent fracture toughness data is illustrated for some fracture results on ABS compounds tested over a range of temperatures. This testing procedure produced quite a wide range of toughness data some of which are valid in terms of compliance to the size conditions, as given in Eq. S9.8.1.4, but some of which were invalid (particularly the data generated at higher test temperatures). All the data are shown in Fig. S9.8.1/1, where ductility factor is plotted against fracture toughness (G_c).

When the data are valid (left-hand section of the plot), there is good correlation, as would be expected. When the data are invalid, there is considerable scatter (right-hand section of the plot). The point that emerges is that clarity is attendant on the valid data but absent from invalid data.

Of course, this does not just apply to fracture mechanics data. When standard impact tests are conducted on notched specimens, particularly when blunt notches are used, it is highly likely that conditions are not appropriate for initiating plane strain cracks. Considerable scatter and an associated ambiguity in the results can therefore be expected, as exemplified in Chapter 8, Supplement S8.6.2.

Reference

1 E. Plati and J. G. Williams. 'The determination of the fracture parameters G_c and J_c for polymers in impact.' *Polymer Engineering Science* 15, 6, 470, 1975.

S9.8.2 Fracture toughness of polypropylene as a function of processing factors

A picture frame cavity with several gating options provides a source of test pieces with various flow lengths and flow ratios, knit-lines and other flow geometry features. Work on this type of mould is described in reference 1, but the fracture and other studies were conducted in a 10-year period prior to the publication of that reference. Some of the unpublished results (by D. R. Moore, C. J. Hooley and M. Whale, 1980) are reported now, where the quality of fracture toughness results can be considered in the context of some processing factors.

Single edge notch flexure specimens of a propylene–ethylene copolymer of MFI = 4, taken at various positions around a picture frame moulding 6 mm thick have enabled fracture toughness (K_c) at $-70\,°C$ to be determined. The procedures for determining fracture toughness with a three point bend configuration are given in Supplement S9.8.1. The position of a test piece from the picture frame mould is identified by the flow path length at the centre of the specimen where the notch was machined. Results for K_c and yield stress are summarized in Table S9.8.2/1.

Table S9.8.2/1 Fracture mechanics data at $-70\,°C$ for a propylene–ethylene copolymer

Flow path length (mm)	Fracture toughness (K_c) ($-70\,°C$, 100 mm/min) ($Mpa\,m^{1/2}$)	Tensile yield strength (23 °C, 5 mm/min) (MPa)
0	3.2	26
130	4.1	26
220	4.1	25
350	4.0	26

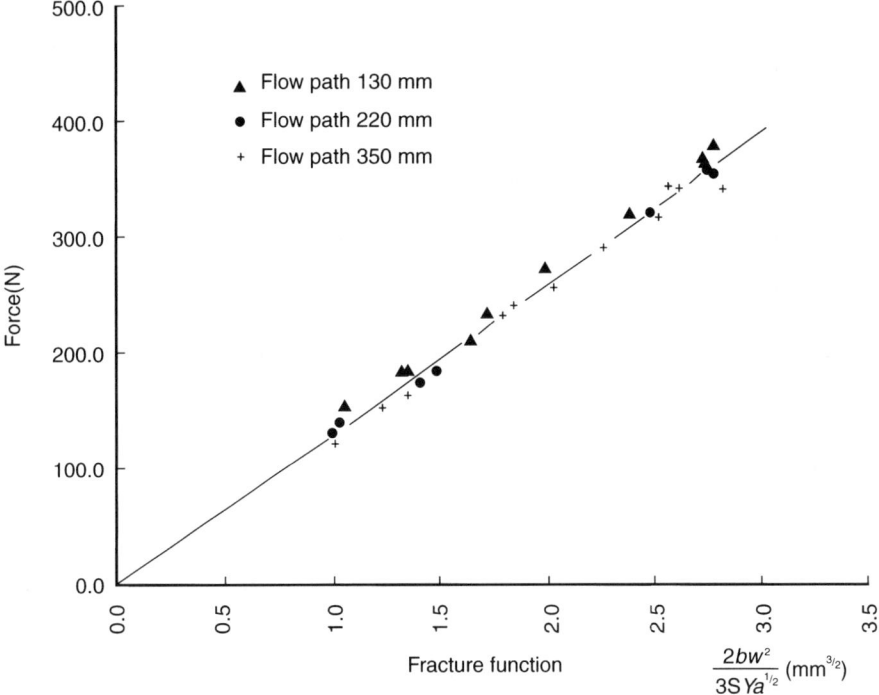

S9.8.2/1. Force vs. fracture function for three different flow path lengths for single edge notched specimens of propylene-ethylene copolymer tested at 100 mm/min at –70 °C for a notch tip radius of 10 μm.

For flow path lengths other than zero, there was no significant difference in mechanical properties for those 6 mm mouldings. Specimens cut from a region adjacent to the core, at zero flow path length, exhibited different mechanical behaviours, which might be explicable in terms of the state of the core material and/or the local orientation. For test pieces taken at flow path lengths in the range 130–350 mm, the yield stress and fracture data indicate a common material state. The fracture mechanics data are plotted in Fig. S9.8.2/1. The fit of the data to a straight-line plot indicates that these fracture data fit a linear elastic theory. The fracture mechanics tests were conducted at a temperature of –70 °C in order to attempt to achieve valid plane strain linear elastic data. However, the tensile yield strength was measured at a slightly slower test speed and at 23 °C, in order to ensure ductility. The ratio of yield strength at –70 °C and at 100 mm/min to that at 23 °C and at 5 mm/min is just larger than 3.[2] Therefore, a notional value for b_{min} is 6 mm (as determined by Eq. S9.8.1.4 in Supplement S9.8.1), which is equal to the specimen width. In addition, a plastic zone size under plane strain conditions at –70 °C is determined to be about 140 μm (compared with a

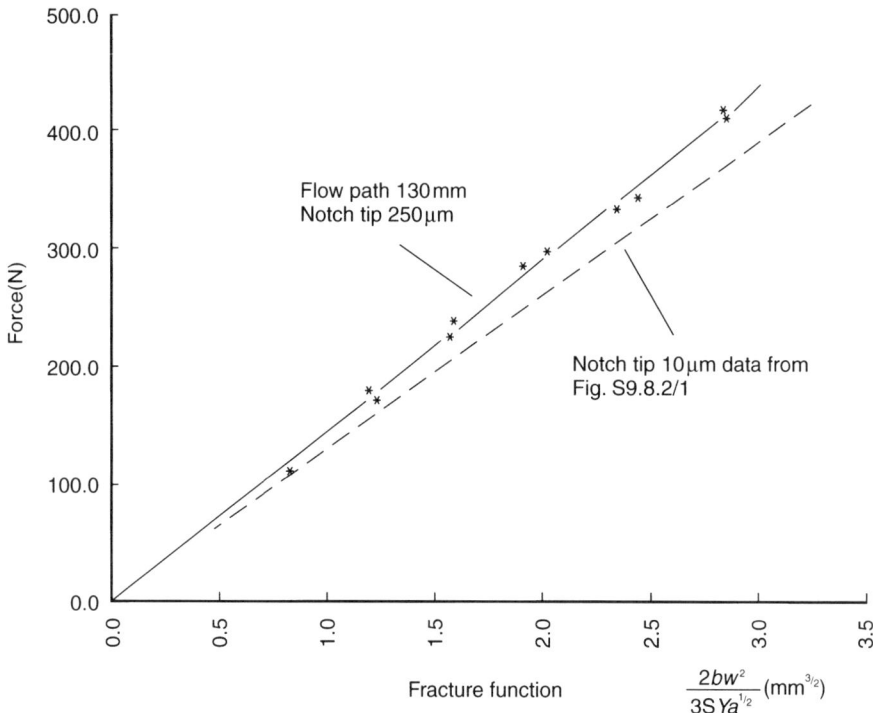

S9.8.2/2. Force vs. fracture function for single edge notched specimens of propylene-ethylene copolymer tested at 100 mm/min at −70 °C for a notch tip radius of 10 mm and 250 μm.

notch tip radius of 10 μm). Consequently, the fracture mechanics tests at −70 °C provide valid plane strain fracture toughness values.

The lower value of critical stress field intensity factor for test pieces cut from a region adjacent to the gate (0 mm flow path length) suggests that pieces cut at flow path lengths 0 mm and 130 mm may be used to generate an anisotropy ratio for stress field intensity factors. In this case, a value of 1.31 is obtained for plane strain conditions.

A notch tip radius of 250 μm (instead of 10 μm) generated more plastic deformation at the tip of the propagating crack. Results from tests at −70 °C are given in Fig. S9.8.2/2. Again, a straight line relationship is observed but with a different value for the critical stress field intensity factor (4.59 MPam$^{1/2}$). On this occasion, the notch tip radius (250 μm) is larger than the plastic zone for the material at −70 °C (140 μm). Consequently, the notch will not act as a sharp crack. Therefore, it can be argued that this value of K_c is probably a plane stress value. Tests at a higher temperature (−50 °C) on specimens with notches of tip radius 10 μm gave data that deviate from a linear relationship when plotted on the same axes as those in Fig. S9.8.2/2, indicating that the fractures were not linear elastic.

References

1 R. A. Chivers, D. R. Moore and P. E. Morton. 'The influence of flow length on the stiffness and toughness of some engineering thermoplastics.' *Plastic Rubber and Composites: Processing and Applications* 15, 3, 145, 1991.
2 J. G. Williams. '*Fracture Mechanics of Polymers*', Ellis Horwood, 1984, p. 154.

S9.11 Tensile yield strength in relation to engineering design

Tensile yield strength (σ_y) is often quoted as a single-valued datum rather than as the multivalued function of temperature, deformation rate, elapsed time and environment that, it is known to be. When it is documented as a function of these variables, it is useful for design calculations where avoidance of failure rather than a precise limit on the magnitude of the strain is the operational criterion.

When it is quoted only as a single value, it can still be used in design calculations if it is modified by an appropriate 'rerating factor' to allow for the effects of temperature, straining rate, etc. Thus, a standard σ_y datum measured at 23 °C on a tensile testing machine at an extension rate of, say, 5 mm/minute might be convertible into other forms of strength:

1   a long-term creep rupture strength by means of a derating factor of ~0.6;
2   a fatigue limit stress by a factor of ~0.1;
3   a short-term 1% isometric stress by a factor of ~0.2.

The actual values of the reduction factors vary with the circumstances, e.g. the ambient temperature, the molecular state of the test piece, etc., see Table 9.3/2 for instance and also Chapter 10.

To a first approximation, the derating factors are multiplicative and hence one yield stress and a few factors could replace the families of creep and creep rupture curves (Chapter 10) at various temperatures, albeit with sacrifice of accuracy. Therefore, in some circumstances even a single-valued yield stress datum may be regarded as a useful engineering design property. It was used as such in the past with judicious application of 'safety factors' but without the refinement of quantitative derating and it could then justifiably be criticized as potentially misleading. Now, as in the case of creep data, see Chapter 6, the adoption of derated yield stress as an engineering design datum reduces the overall cost of data generation by a substantial factor and simplifies the choice of (notional) property values for use in design calculations.

Additionally, the tensile yield behaviour of many plastics is superficially similar to that of some competing engineering materials, e.g. light alloys, even though the physical mechanisms are different. Therefore, comparisons between different classes of material on the basis of that property can be helpful, provided allowance is made for interclass differences in the sensitivities of the yield stress to temperature, straining rate, stress field, etc.

S9.12 Ramp excitation of centrally loaded, edge-supported discs

The results from conventional strength tests in tension, flexure, or shear, may give an over-optimistic picture of the likely service behaviour of end-products. In part, this is because of the relatively simple stress fields that are entailed with standard laboratory test configurations and also because of atypical states of order in test samples, as has been stated elsewhere in this monograph. A possible corrective strategy that has been discussed in various contexts elsewhere in the monograph is the direct testing of end-products. In the particular case of strength testing that raises various issues such as:

1   the choice of test configurations for complex shaped items;
2   the infinite variety of shapes in service;
3   the obscure link between breaking force and breaking stress for all but the simplest shapes;
4   the high breaking forces entailed if the load-bearing cross-sections are large;
5   the effect of shape on the plane stress to plane strain transition.

A compromise strategy that circumvents some of those issues is the testing of simple structures. These are intermediate in shape between idealized test specimens and service items. Some of these have been discussed earlier, i.e. critical basic shapes (see Chapter 1, Supplement S1.11), one of which is a square plaque that can be variously gated. Discs cut from such plaques or discs moulded directly are tested in three-line flexure for the assessment of flow-induced modulus anisotropy, see Chapter 7, Supplement S7.7. In principle, centrally loaded, edge-supported discs should yield a mean value of the in-plane modulus, but in practice the procedure is unreliable because the attainable precision at the necessarily small strain is low. On the other hand, that configuration is commonly used in impact testing, see Chapter 11. It has been very informative also at lower deformation rates in ramp excitation tests[1] because the progressive development of damage can be viewed as the test progresses without resort to the more elaborate analysis methods needed for impact tests. For example, it can identify important differences between the failure characteristics of two commercial grades of polypropylene, each containing 0.20 weight fraction of short glass fibres. One grade contains the fibres chemically coupled to the matrix and the other having no such coupling (the former is described as a 'fibre-reinforced' grade and the other is described as a 'fibre-filled' grade). The materials were injection moulded into double-feed weir-gated plaques 150 mm × 150 mm × 3 mm in size, see also Chapter 7, Supplement S7.7 and Chapter 2, Supplement S2.7. Discs approximately 150 mm in diameter were cut from the plaques, supported at their periphery and flexed by an actuator with a hemispherical tip of radius 50 mm acting at the centre point of the disc and moving at a rate of 10 mm/min.

Discs of the uncoupled grade deformed so much that they were pulled through the support ring, whereas discs of the coupled grade failed by high energy crack propagation. Comparisons such as that vary with the test con-

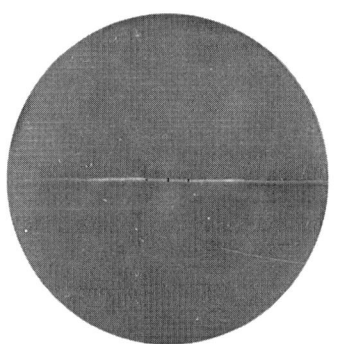

Reinforced grade
Notch along knit-line

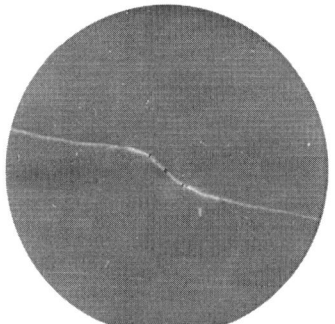

Reinforced grade
Notch at 45° to knit-line

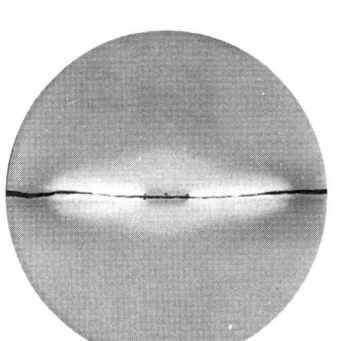

Filled grade
Notch along knit-line

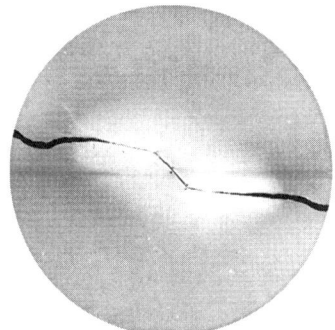

Filled grade
Notch at 45° to knit-line

FAILURE OF DISCS WITH SURFACE NOTCHES
Polypropylene containing glass fibres

S9.12/1. Centre-loaded, edge-supported discs of polypropylene containing 0.20 weight fraction of short glass fibres. Surface notches along and at 45° to the knit-line.
In the uncoupled grade the growth of initial crack is restricted by a pseudoplastic zone caused by phase separation.

ditions; at the straining rates entailed in impact tests both grades would be brittle, but differently so, and the two grades are differently sensitive to notches. The effect of a notch in an anisotropic test piece depends on the alignment of the notch in relation to the anisotropy axes and interesting effects can be seen when the alignment of a shallow surface notch with the profile of an arc is varied. Figure S9.12/1 shows how the cracks initiated by notches along the knit-line and at 45° to it are subsequently influenced by the flow-induced anisotropy. The knit-line arising from the double-feed flow geometry is clearly visible. The cracks obviously grew initially in the direction dictated by the notch axis, but subsequently propagated along the axis of strength anisotropy associated locally with the knit-line. The white regions in the photographs of the discs moulded from the uncoupled grade show where the glass fibres have separated from the matrix. Crack growth is inhibited by the phase separation, which redistributes the stress field at the crack tips and acts in a similar way to plastic deformation. The same phenomenon is often seen in other fibre-reinforced composites.

With notches at 90° to the knit-line (not shown), the crack propagated unhindered along the direction of the notch in the coupled grade and was stopped after little growth by phase separation. The numerical results are summarized in Table S9.12/1. They are fully consistent with the observed behaviour of the cracks.

The uncoupled material is undoubtedly the tougher of the two, and such grades are promoted commercially as such, but the coupled grade could not be described as weak because its crack initiation energy was high. Similar conclusions can be drawn from standard tensile tests but the course of the

Table S9.12/1 Failure data for centre-loaded, edge-supported discs injection moulded from polypropylene containing 0.20 weight fraction of short glass fibres

Material	Notch alignment[a]	Maximum force (N)	Failure energy (J)
Coupled grade	Unnotched	3340	27.8
	0° notch	872	4.26
	45° notch	1280	5.79
	90° notch	1420	7.56
Uncoupled grade	Unnotched	Not valid	Not valid
	0° notch	1130	36.9
	45° notch	2000	42.7
	90° notch	>3500	>50

[a] with respect to the knit-line.

fracture processes are less clearly demonstrated by the response curves; in this case, as in many others, several tests would have been required to identify and quantify the anisotropy.

Reference

1 C. M. R. Dunn and M. J. Williams. 'Measurement of the strength of plastic plates.' *Plastics and Rubber: Materials and Applications* 5, 90, 1980.

10

Strength and ductility from step-function and cyclic excitations

10.1

Strength and ductility generally relate to phenomena occurring at relatively large strains for which the formal theory of linear viscoelasticity is unhelpful. However, for reasons of practical expediency, the same excitation functions are used for most of the experimentation.[1] Step excitations are appropriate for the study of creep rupture lifetimes and the phenomena of damage development under sustained stress. Stress (or strain) cycling or non-periodic intermittency is separately employed to study fatigue strength where premature failure and/or embrittlement is induced. For the latter class of experiment, periodic square waves are superior to sinusoids on analytical grounds when the strains lie outside the linear region but even so sinusoids are often preferred on practical grounds.

The traditional short-term strength tests for plastics cannot characterize the long-term behaviour, because they impose failure via a continuously increasing strain and thereby override certain slow deterioration processes that would otherwise have been manifest during protracted periods under stress. Similarly, step excitation tests of short duration are unlikely to give any information about very slow failure processes.

The characterization of long-term durability requires multipoint data covering the effects of elapsed time under load, temperature, stress field, stress history and other variables, for which there are few international standard test methods.

10.2

The preferred deformation mode for creep rupture studies is now tension. Flexure was once popular and is still resorted to if the material has a very high modulus and in certain special cases such as environmental stress cracking tests, where it is often necessary to induce rupture in intrinsically tough grades of a relatively low modulus material, see Sections 10.11 and 10.12 and Supplement S10.11.

There is no unique criterion of durability for plastics in load-bearing situations; there may be various events during the lifetime of a plastic item, e.g. the development of crazing or stress-whitening, which do not necessarily impair its usefulness for load-bearing, but which may reduce its usefulness in other ways. Furthermore, a crack may be tolerable if its rate of growth is so slow that the expected time to complete failure comfortably exceeds the lifetime requirement of the application.

Failure in polymer–fibre composites is a similarly progressive matter, but more complicated because of the dissimilarity of the mechanical properties of the two main constituents; failure may be gross fracture, resin fracture or phase separation.

10.3

Creep rupture is the termination of creep by failure, where a wide variety of failure types may have to be identified, accommodated and possibly adopted as a failure criterion. Even 'complete rupture' is not quite as definitive as it might seem at first sight. For convenience within the framework of long-term durability studies, however, creep rupture is usually regarded as either the onset of yielding or fracture without gross yielding, but both the onset of yielding and the development of a crack of critical length can be protracted and ill-defined processes that cannot therefore quantify the lifetime precisely.

Thus, the experimental lifetimes derived in creep rupture tests are often arbitrarily artificial; at its simplest a creep rupture datum is the time at which rupture occurs under a particular sustained stress and a creep rupture lifetime curve is the best curve fitting sets of data obtained at several stress levels, though it is arguable as to what the curve-fitting criterion shoud be; for instance, should it be the mean log lifetime or the shortest lifetime in each set of data and how large should the set be? The shape of such curves and the magnitude of rupture strength varies with temperature, see Fig. 10.3/1, for example, which shows idealized rupture lifetime curves of PMMA at several temperatures. They define the upper bound of the load-bearing capability of the material at those temperatures.

Ostensibly, a ductile failure lifetime curve denotes the time to yield under various stresses but the nature of the yielding in thermoplastics usually changes along the curve. Failures at longer times are less ductile than those at shorter times and with limited development of a plastically yielded zone. There are well-authenticated instances of the failures at long times being unambiguously brittle with an abrupt transition from ductile failures at short lifetimes to brittle ones, but the more usual behaviour is a gradual ductile–brittle transition as shown in Fig. 10.3/2. That figure shows, addi-

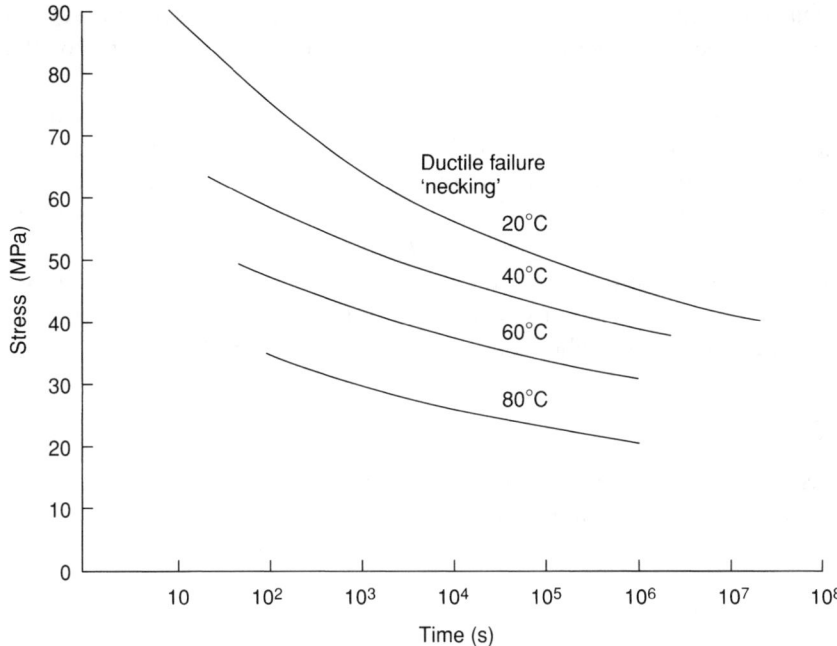

10.3/1. Creep rupture of PMMA at various temperatures. Creep rupture is defined here as the onset of necking.

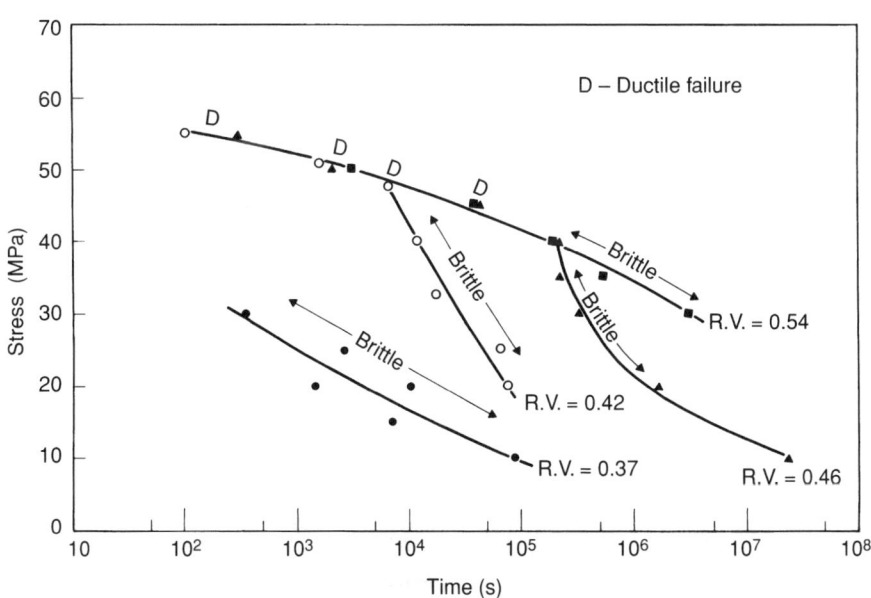

10.3/2. Ductile–brittle transition in the creep rupture of specimens of poly(ether sulphone). The abruptness of the transition and its position on the time axis vary with the molecular weight of the polymer.

tionally, that the position on the time axis of the transition from predominantly ductile behaviour to predominantly brittle behaviour depends on the average molecular weight of the polymer.

The curved form of the ductile failure section and the possibility of an increased downward trend of the lifetime curve at longer times limits the likely reliability of any extrapolation process. However, irrespective of whether the change in failure mode is abrupt or gradual, the time at which the transition occurs decreases with increasing temperature and therefore the transition point can be taken as the reference marker for time-temperature superposition into a creep rupture 'master curve'[2] (see Supplement S6.6 for its application to stress relaxation data).

10.4

It is common practice for creep rupture data to be presented in conjunction with the corresponding creep data in the form of isometric curves, i.e. sections at constant strain across a family of creep curves. Figure 10.4/1 is

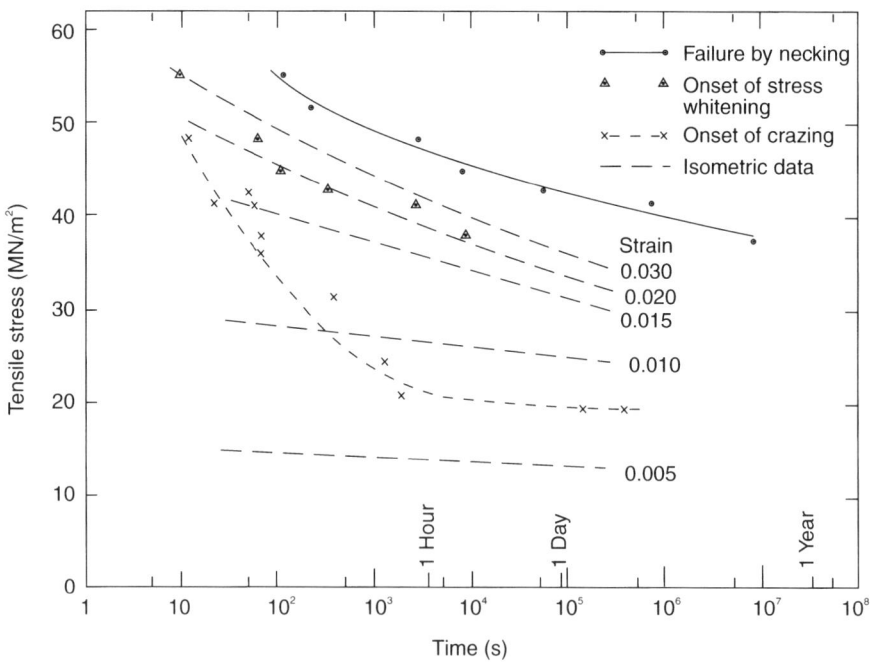

10.4/1. Creep rupture data combined with creep isometric curves. PVC at 20 °C and 65% r.h. (from reference 1).

an example. The ductile failure curve usually follows the isometrics in general form, which implies that the strain to yield is approximately constant, though there is some evidence of lowered yield strain at the longer lifetimes, corresponding to the reducing ductility referred to in Section 10.3. If failure is defined by some other criterion, e.g. the onset of crazing, that failure curve cuts across the isometrics, reflecting the known complexities of craze development.

Stress concentrators induce brittleness, mainly via stress field triaxiality and reduce lifetimes correspondingly. Aggressive environments have similar overall effects but via stress-enhanced disruptive processes. The effects of both stress concentrators and aggressive environments can be shown on the same type of diagram.

10.5

The apparatus that is employed for creep rupture tests on flat, simple test pieces is similar to, but simpler than, that used for creep studies. Since the main interest lies in lifetimes and the various manifestations of failure, there is no need for the early part of the deformation response to be known precisely and hence no dominating requirement for rapid loading; therefore a need for subsidiary loading mechanisms and accurate strain measurement is removed. Correspondingly, there is a reduced requirement for accurately axial loading in tension tests and the less demanding criteria can be satisfied via long and flexible linkages. That also allows several test pieces to be tested in series on one machine, with the obvious advantages of lower capital investment and timely identification of the earliest failure in a set of test pieces but the disadvantage of no information on lifetime variation. On the other hand, it is desirable that the onset of damage should be observable, which requires good visual access and a specimen shape that limits the region in which failure and prefailure phenomena are likely to occur, i.e. a waisted specimen or a notched one, even though that changes the local stress field. Various notch tip radii are used, fairly standard values being 2 mm, 1 mm, 500 μm, 250 μm and 10 μm. Fracture mechanics analysis is increasingly used but there is some uncertainty over its accuracy when the crack grows very slowly because K_{I_c} is not constant over long lifetimes; Burn[3] has recommended that the short-term value should be linearly extrapolated along a slope of approximately $-0.3\,\text{MPa}\sqrt{\text{m}}$ per log decade.

Thermoplastics test pieces can be directly moulded or cut from larger items as opportunity dictates or occasion demands. It is usual for test pieces of thermosetting materials to be machined from flat panels and the edges polished if the material is brittle. Test specimens may even be end-products. Extruded tubes subjected to internal pressure are one such special case; the strength in the hoop direction is often the critical issue but combinations of internal pressure and end-constraints establish biaxial stress fields, which are sometimes studied in a search for a general yield criterion, but more often for the prediction of rupture and fracture lifetimes in pipelines for the conveyance of gases and liquids. Various standard tests stipulate procedures that are intended to check the quality of such classes of extruded pipe; some entail the extrapolation of short-term data to 50 years, for which time a required minimum hoop breaking stress is postulated. Those test programmes can become extensive and expensive because the many factors that can affect pipe quality have to be taken into account. It is then sometimes useful if rings cut from the pipes, i.e. very short sections of pipe, are tensioned along a diameter or via D-shaped plates which subject that part of the pipe ring in the gap between the Ds to combined tension and bending;[4] the gauge sections of that test piece may be waisted or notched as required.

ASTM D2990-77 (1982), *Tensile, Compressive and Flexural Creep and Creep Rupture of Plastics* offers some general guidance on test procedures. Creep rupture data (and fatigue data) are notoriously variable; a set of specimens tested at one stress may have a range of lifetimes such that $t_{max} = 50t_{min}$. At first sight that may seem so excessive as to indicate poor experimentation or variable quality of the test pieces, and it is certainly an experimental inconvenience, but the slope of the ductile section of a log lifetime curve is often so shallow that a coefficient of variation of 0.03, which is typical for ductile failure stress in tension, translates into a wide band along the log lifetime axis. That scatter poses difficulties for the experimenter and compromises the extrapolation procedures, especially when the test programme relates to the prospective lifetime of a major installation such as a pipeline.

All lifetime tests are beset by the fact that the failure may be premature because of extraneous deficiencies in the material or deficiencies in the processing and in those cases where the datum is suspect, as an outlier for instance, the initiation site and the progression path of the failure should be inspected for evidence of abnormalities, but there should be great restraint over the rejection of a datum.

10.6

Creep rupture data that embrace anisotropy, stress concentrations, flow geometry, etc., all of which affect lifetimes, define allowable and forbidden regions of stressing but they can be complicated. In that case, the use of derating factors can simplify the presentation and the utilization of the data. For instance, the ductile creep rupture stress at short times is not very different from the value of yield stress derived in a conventional stress–strain

Table 10.6/1 Creep rupture derating factors for nylon 66 + 0.33 weight fraction of glass fibres at 23 °C

Test specimen	Lifetimes (s)	Rupture stress (MPa)	Derating factor
Standard bar	Short	187[a]	1
Waisted specimen cut			
from standard bar	10^2	162	0.86
ditto	10^7	121	0.64
Sharply notched specimen cut			
transverse to flow direction	10^2	69	0.36
ditto	10^7	46	0.24

[a] This was the value obtained in a particular experiment. The corresponding value quoted in Table 9.3/1, which is for the same property, was abstracted from the manufacturer's data sheet.

test on a specimen of the same geometry, and that latter datum can therefore be used as the upper bound or reference value. Stress levels for longer lifetimes can then be expressed as decimal fractions of that upper bound value and hence, with some sacrifice of detail, a complicated diagram can be reduced to a simple tabulation, e.g. Table 10.6/1.

10.7

In service, the stress is often intermittent or cyclic and the lifetime is then usually less than under steady stress, i.e. plastics, like metals, are prone to fatigue. Fatigue lifetimes vary with stress mode(s), waveform and the symmetry of the cycle about zero stress or strain. Consequently, a comprehensive evaluation of fatigue resistance requires an expansive test programme and/or a test strategy analogous to that necessary for economic evaluation of creep behaviour (see Chapter 6, Supplement S6.14).

Fatigue in metals is normally studied via sinusoidal stressing or straining, and that practice has been followed for most of the fatigue tests on continuous fibre reinforced composites and for many of those on plastics. The mechanical arrangements for sinusoidal stressing or straining are relatively simple. However, it is often desirable that creep rupture data and fatigue data for plastics should be combined on a single diagram, but the stress axes cannot be easily equated if the fatigue excitation is a sinusoid and/or the strains are above about 0.005. A repeated gate function as the excitation leads to simple correspondence and; additionally, a logical link between intermittent and periodic excitations, see Section 10.9 for additional notes on that, and see Supplement S10.7 for notes on experimental methods.

10.8

The lifetimes of plastics under fluctuating stress are strongly influenced by their viscoelastic (and generally non-linear) nature. This is less critically the case for metals. However, it has been important in contemplating plastics for engineering applications to give consideration to the style of presentation of data that has become established for metals. Metals fatigue data are usually presented as stress vs. log number of cycles to failure, designated in this context as S vs. $\log N$. For the majority of metals the stress decreases gradually as $\log N$ increases, with an endurance limit or fatigue lifetime in the region 10^7–10^8 cycles. For ferrous metals the S/N curve flattens out to reveal a definite 'fatigue limit', a value of the stress amplitude below which the material may survive an infinite number of cycles. That fatigue limit is far less obvious for plastics.

The principal differences between fatigue in metals and plastics are summarized in Table 10.8/1. Those differences are reflected in the procedures and strategies commonly used for fatigue tests on metals and plastics, respectively.

Table 10.8/1 Characteristic Features of the Fatigue of Metals and Plastics

Influential Variable	Metals	Plastics
Frequency	Stresses are generally in the region of or below the elastic limit, so that hysteresis and associated heating is slight and frequencies up to 150 Hz have little influence on fatigue behaviour. The S/N curve can therefore be extended to values of N–10^8 in a relatively short experiment.	High mechanical hysteresis and poor thermal conductivity can lead to marked rises in temperature. Failure of specimens may then be by thermal softening. To avoid that in test programmes, much lower frequencies have to be used, e.g. 0.1–10 Hz. Test periods may thus be protracted.
Mode of deformation (flexure and uniaxial tension)	There is little if any difference between results obtained in flexural and uniaxial tension tests.	In view of the different levels of strain energy involved in the two test configurations there may be an effect. There is little evidence about correlations between results from the two configurations.
Level of the mean load.	At high temperatures and at stresses in excess of the yield stress at room temperature a non-zero mean stress causes creep	A non-zero mean stress gives rise to accumulative creep as N increases. This affects the fatigue behaviour

10.9

Any intermittency of the stressing of plastics reduces the lifetime (expressed as time under stress) and reduces the ductility. The use of a square stress waveform (frequency f), with equal times on and off load, provides a simple analysis since the total time on (and off) load is given by $t_{ON} = \dfrac{N}{2f}$. If the waveform is a sinusoid or some other form, then the transformation of frequency to time on load is a more complex function. The severity of this effect is a function of the duration of the zero load or reduced load periods. It seems that the better the recovery during such periods, the poorer the endurance and the more 'brittle' the failure; thus, for example, glassy amorphous plastics are more seriously affected than crystalline plastics above their glass–rubber transition because they recover more promptly after creep (though that statement is an oversimplification of recovery behaviour, see Chapter 6).

10.10

Thermal failure, which is referred to in Table 10.8/1, is the softening/collapse of the test piece due to a substantial rise in temperature that may occur during prolonged stress/strain cycling at frequencies higher that a few hertz. Smith and Sauer[5] reported temperature rises of a few degrees in air-equilibrated PMMA test pieces (i.e. what would be regarded as a normal PMMA sample) fatigued in tension–compression at a frequency of only 2 Hz and a rise of 42 °C in a sample containing 2.25% water. Internal energy dissipation is an important limitation in the fatigue testing of plastics because it excludes the metals testing technique of using high frequencies to shorten fatigue lifetimes and thereby the test duration. This can also be the situation in the testing of continuous carbon fibre reinforced thermoplastics, as discussed in Supplement S10.10. It arises because the response lags behind the excitation in a viscoelastic test piece and the heat generated depends on the magnitude of the phase difference at the frequency and temperature of the test. Also, if a small increase in the specimen temperature effectively moves the test conditions up the rising flank of a complex modulus loss process, the experimental situation is potentially unstable. The size of the test piece and factors such as external heat dissipation processes affect the propensity to thermal failure, which is usually manifest as a loss of form stability.

Thermal failure also intervenes to reduce fatigue endurance in service but, because of the likely disparities between the physical details of a fatigue test and those of a service situation, correlations between such failures in a laboratory and in service are tenuous. Therefore, thermal failure in a fatigue test should be regarded both as a test malfunction and as an indicator of inappropriate test conditions.

10.11

Creep rupture tests are sometimes carried out in aggressive environments and are then referred to as 'stress corrosion' or 'environmental stress cracking' tests. The latter were originally accorded a special cachet, at least partly because they impinged on a demanding yet lucrative application (insulation in transoceanic telephone cables), but the phenomenon is of more general importance because certain environments are selectively aggressive towards particular polymers and can reduce the load-bearing lifetime, and sometimes even the zero-stress storage lifetime, significantly relative to that in air, see Supplement S10.11 for some details. Figure 3 of reference 6 is a good illustration of the consequences of attack by an aggressive fluid on a polymer that is widely and properly regarded as tough in most circumstances.

In certain circumstances a fracture mechanics analysis may be used to calculate safe working stresses or likely lifetimes in aggressive media, but the legitimacy of that process has to be assessed in the light of the nature of the interaction that occurs between the material and the environment.

10.12

The failure-inducing stress or strain may be an externally applied one, an internal one that has arisen as a consequence of the processing stage or a combination of the two. Internal stresses and, indirectly, the 'quality' of a moulding may be gauged by a 'solvent-dip' test in which the moulding or a part cut from it is immersed in a solvent known to be mildly aggressive towards the material. Figure 10.12/1 shows cracks lying along the lines of molecular orientation in the surface layers of two injection mouldings (depending on the thickness of the moulding and the processing conditions, the molecular alignment in the core of the mouldings may be mainly orthogonal to that in the surface layers). Thus, immersion in a liquid that is only mildly aggressive can be helpful on 'flow visualization' and also as an indicator of a likelihood of future incidence of so-called 'after-cracking', i.e. the delayed cracking of nominally unstressed mouldings due to residual stresses that originated during the manufacturing process.

10.13

The damage induced in the solvent dip and in the more quantitative environmental stress cracking tests range from slight crazing to severe cracking

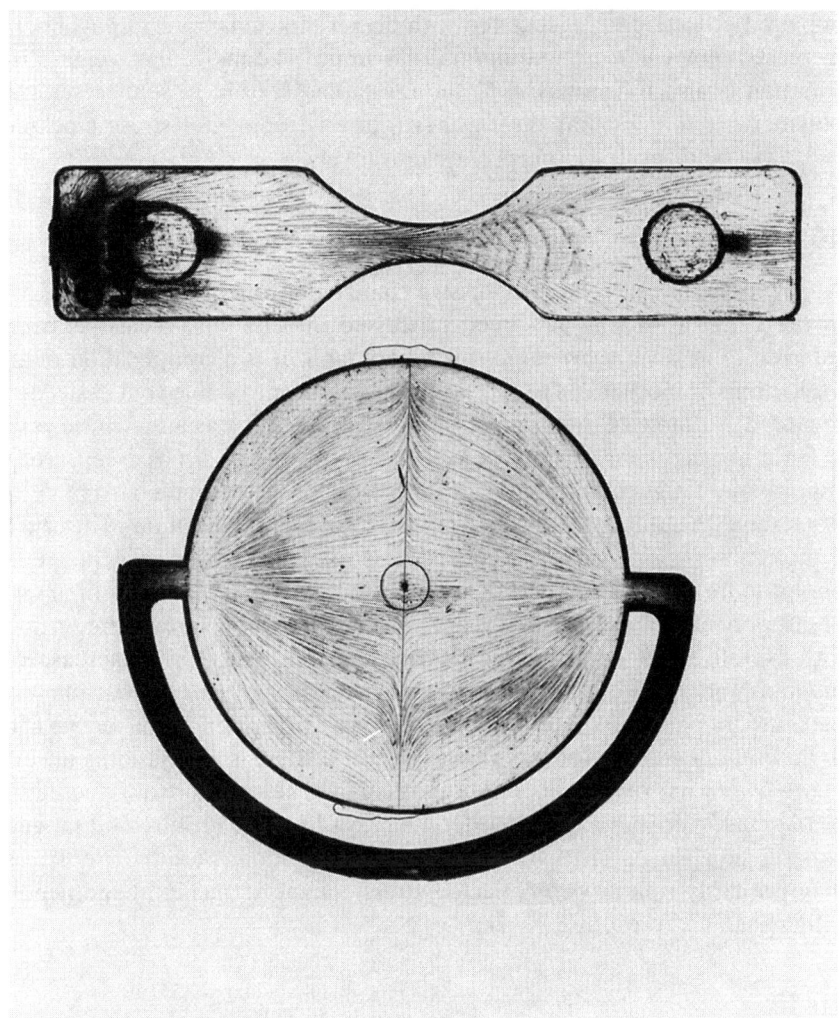

10.12/1. Cracks induced in injection mouldings by immersion in a mildly aggressive liquid. In such cases, the stress is residual stress arising from the moulding operation. The cracks reveal the pattern of molecular orientation in the surface layers. The severity of damage depends on the environment and the duration of immersion.

and from an instantaneous response to an almost imperceptiby slow progressive deterioration. As a general rule, cross-linked polymers are more resistant to environmental stress cracking than uncross-linked ones and crystalline polymers are more resistant than amorphous ones. Also, to a first appproximation, the higher the molecular weight, the greater the resistance, but for injection mouldings that trend often reverses at some point as the impaired processability associated with higher molecular weight engenders increased molecular orientation/internal strain. The molecular weight distribution is also influential, as is the crystalline texture in semicrystalline polymers. It follows that the results from environmental stress cracking tests may be difficult to interpret quantitatively.

10.14

Ageing as manifest as the progressive change in properties that take place in the material after it has been processed into its final form has been referred to at several points in this monograph. It is a complication in all evaluations of mechanical properties, either as a controllable (but easily disregarded) variable of the pre-test storage conditions or as an intrinsic variable during the measurement of long-term properties such as creep, creep rupture and fatigue. Allowance for its effects should be made also in decisions about the utilization of materials over long service lifetimes. The cause is molecular rearrangement, which is sometimes observable as a change in one or more of the structural parameters, but which, particularly in amorphous polymers, may only be inferable from changes in certain properties. The overall effect on mechanical properties is a progressive increase of modulus and a loss of ductility with increased storage time so that the test pieces that contribute datum points to an experimental lifetime curve age to different degrees before they fail. Therefore, there is always some uncertainty over what the results imply, particularly the change from ductile to brittle characteristics along the lifetime curve in creep rupture and fatigue experiments. The embrittlement and failure after a long period under stress may primarily reflect a complicated stress-enhanced ageing phenomenon rather than creep rupture *per se*.

10.15

The effect of a combination of unfavourable flow geometry, stress concentrator, intermittent stress and aggressive environment can be devastating for thermoplastics. In creep rupture and fatigue investigations those behavioural features can be used in two ways: as accelerating devices to shorten lifetimes and/or to induce early brittleness which has the same overall effect on the lifetime. An illustration of accelerated experimentation is given in Supplement S10.15 in generating design data for the load-bearing design of polyethylene water tanks. However, though there are many such data, the quantitative reliability of inferences drawn from them about 'normal' creep rupture lifetimes has not been established and their main role should remain as providing worst-case lifetime data rather than as predicted upper bound creep rupture data.

Thermosets and composites are correspondingly sensitive to factors such as state of cure, fibre alignment, voids, etc. in addition to the mechanical variables such as stress level, waveform and environment. Where the service requirements are extremely exacting, as in aeronautical applications for instance, there is great reliance on special fitness-for-purpose tests.

References

1 K. V. Gotham. 'A formalized experimental approach to the fatigue of thermoplastics.' *Plastics and Polymers*, 37, 309, 1969.
2 F. R. Larsen and J. Miller. 'A time–temperature relationship for rupture and creep stresses.' *Transactions of the American Society of Mechanical Engineers*, 74, 765, 1952.
3 L. S. Burn. 'Lifetime prediction of uPVC pipes – experimental and theoretical comparisons.' *Plastics, Rubber and Composites: Processing and Applications*, 21, 2, 99, 1994.
4 D. R. Moore, P. P. Benham, K. V. Gotham, M. J. Hitch and M. J. Littlewood. 'Long-term fracture performance of a uPVC pressure pipe as influenced by processing.' *Plastics and Rubber: Materials and Applications*, 5, 4, 146, 1980.
5 L. S. A. Smith and J. A. Sauer. 'Sorbed water and mechanical behaviour of poly(methyl methacrylate).' *Plastics and Rubber: Processing and Applications*, 6, 1, 57, 1986.
6 K. V. Gotham and D. C. Wright. 'Fatigue in polycarbonate.' *Plastics and Rubber: Processing and Applications*, 4, 1, 43, 1984.

S10.5 Practical aspects of tensile creep rupture testing

Tensile creep rupture equipment is not usually built to the exacting specification that is essential for high grade creep machines, because the traditional form of creep rupture testing measures lifetime as a function of applied stress and the strain does not have to be measured and is not measured except in special experiments. Additionally, a cost-limiting strategy has to be adopted since the inherently high interspecimen variability characteristic of the tests necessitates the use of numerous test pieces.

In the case of thermoplastics, direct injection moulding of tensile creep rupture test pieces should be avoided because the molecular orientation in such pieces (usually end-gated) can have a misleadingly enhancing effect on the lifetime. Test pieces should be machined from plaques or other shapes and their position and alignment with respect to the complete moulding should be regarded as additional test variables.

The high interspecimen variability creates some uncertainty about the quality of creep rupture data. The datum that is chosen as representative of the lifetime at a particular level of applied stress could be the mean value of a set of lifetime data from replicate tests on nominally identical specimens, but ideally should perhaps be the shortest rupture time in such a set. Apart from considerations of the precision, accuracy and physical significence of each such datum, the investigator needs to know the goodness of fit of the regression line, the reliability of any extrapolation and the implications for any derived estimates of safe working stresses.

In the light of that overall requirement, the choice of excitation levels and the number of test pieces at each depends to some degree on the purpose of the test programme. Thus, if the shape of the applied stress–lifetime relationship is required test pieces have to be expended at several stress levels, but possibly concentrated in a region suspected of being, or known to be, a critical one; two levels may suffice if a straight line can be used as an approximation to the actual relationship in a non-critical region. However, the possible consequences of service failure and the cost penalties of malfunction have to be taken into consideration and various statistical procedures are imposed on quality assurance and quality control procedures to safeguard the service performance of some end-products. Various abbreviated procedures have also been adopted for exploratory studies. One such,[1] chooses the first (and highest) test stress in relation to the conventional yield strength (ramp excitation) and the next lower stress on the assumption that the first measured lifetime was the 'true' value; the third stress level is chosen by linear extrapolation of the first two points plotted as stress vs. log lifetime. The fourth level is chosen by linear extrapolation of the best straight-line fit to the first three points and so on until the stress likely to give failures in the range $10^6 s \rightarrow 10^7 s$ can be estimated, whereupon several specimens are tested at that stress. There are, of course, numerous variants of that procedure. Dunn et al.[2] have enhanced the method as applied to fibre composites and introduced the term 'frugal testing'[3] and procedures pertinent to the pipe manufacture, installation and utilization industries are widely disseminated at regular conferences and in special issues of certain technical journals, see reference 4, for instance.

Other additional tests such as creep rupture of notched specimens, fatigue of unnotched specimens and possibly fatigue of notched specimens, can be used judiciously to shorten lifetimes, either to supplement the data cover or to reduce the duration of the 'regular' tests.

None of the advocates of abbreviated testing suggest that the procedures dispense with the need to conduct tests of long duration and the testing strategies all rely on judicious combinations of long-duration and abbreviated tests. The duration of the former tests should ideally be for the 'natural lifetime' of the material, if that is a sensible concept, and in some instances test pieces that survive may be more informative than those that fail. Where new materials or new grades of existing materials are being assessed for important areas of application, well-documented samples should be allocated for long-duration survival tests. It seems, in the light of current experience, that some of the test pieces should be under sustained stresses at between 10% and 25% of the conventional yield stress, some should be subjected to diurnal square wave cycling (say 8 hours under stress and 16 hours unstressed) and some should be stressed under environmental conditions related to the prospective service, e.g. outdoor exposure, in water, or in natural gas.

The numerous procedures, normal or accelerated, which are not, of course, limited to the tensile mode of deformation, all lead to estimates of allowable service stress, but no matter how elaborate or exhaustive the experimental plan or how sophisticated the statistical analysis, maximum safe working stresses will always be beyond totally certain experimental verification. It is worth noting that Dunn and Morgan,[2] 'given the inevitable imperfections in the data', chose to use a statistical analysis procedure that 'provided the proper balance between the mathematics and the judgment of the experimentalist'.

References

1 K. V. Gotham and S. Turner. 'Procedures for the evaluation of long-term strength of plastics and some results for polyvinyl chloride.' *Polymer Engineering and Science*, 13, 2, 113, 1973.
2 C. M. R. Dunn and R. J. Morgan. 'Prediction of durability through statistical distributions.' Proc. 1995 SEM Spring Conference and Exhibit. June 12–14, 1995, Grand Rapids, Michigan, USA.
3 C. M. R. Dunn, E. J. Fouch and R. J. Morgan. 'Frugal testing for auto and aero industries. A novel approach to data resourcing.' Proc. SEM Conference, Baltimore, pp. 382–391, 1994.
4 Various authors. *Plastics, Rubber and Composites: Processing and Applications*, 25, 4, 1996; 25, 6, 1996; 27, 9, 1998; 27, 10, 1998.

S10.7 Fatigue – experimental methods

There are two broad classes of fatigue experiment: First, an approach which explores the relationship between applied stress and the number of fatigue cycles (*S–N* curves); second, a more particular approach for sharply notched specimens where crack growth in fatigue is examined (Paris law type experiments). Flexure and tensile deformation have been regularly adopted for generating *S–N* curves in fatigue. Flexure largely avoids clamping problems and even flexed cantilevers can be used without serious risk of failures at the grips if the profile of the beam is so chosen that the maximum stress occurs some distance from them. For instance, ASTM D671 (*Tests for Repeated Flexural Stress of Plastics*) specifies a linear taper on the cantilever, i.e. a region of uniform stress. The cantilever configuration is more convenient than three-point or four-point bending for fully reversed cycles. A particularly simple device which is popular in metals testing but not widely used for plastics is the so-called 'rotating-bend' machine, in which a cantilever of circular, though not necessarily constant, cross-section is loaded at its free end and rotated.

The basic requirements for a tensile fatigue machine are the same as those for a tensile creep machine, viz. an axial stress system that is accurately prescribed, plus a set of defined and adjustable excitation functions. A range of excitations (sinusoidal, square wave, saw tooth) are readily achieved by modern machines. Axiality is less readily obtained; the method used in creep testing is unsuitable for fatigue experiments because the pin-and-hole coupling entails a stress concentration that is sufficiently severe to induce premature failure in many cases. In general, special grips are needed and the test piece has to be gripped over a relatively large area, to the detriment of the alignment. Owen, amongst others, described a tension fatigue testing machine in detail.[1] He stated that a stress concentration factor of at least 1.05 usually has to be tolerated; he has used a range of material-specific test piece geometries.

A wide range of test piece shapes and sizes are used in tension–compression fatigue testing, but most are similar in general shape to that used in creep rupture tests. If the test piece is to be under uniaxial compression during all or part of the cycle, it has to be squat for stability against buckling, whereas it should be long and slender to facilitate axial stressing. The shape that was adopted by Crawford and Benham,[2] for instance, was a relatively thick-walled hollow cylinder with conical ends. That shape was a practical compromise between the conflicting dimensional requirements, but the research programme also sought to quantify the effects of surface finish, knit-lines and other factors on the fatigue endurance of injection mouldings and therefore the specimen had to be a further compromise between what was required by the mechanics of the experiment and what was a reasonable representation of moulded products. The dimensional and shape requirements for test pieces of continuous fibre composites entail the use of elaborate devices to achieve axial alignment and secure gripping during repeated stress cycles.

Meticulous preparation of the test piece is necessary for fatigue studies. ASTM D671, which summarizes the best practice, stipulates the use of a very sharp cutting tool and a judicious combination of speed and feed to give a good finish with minimum heating, followed by polishing with progressively finer emery papers to remove all surface marks from tools, etc. Final polishing has to be lengthwise and all the polishing has to be either by hand or with light pressure and a slowly revolving drum. The edges and corners of the test piece should not be rounded. With such requirements for painstaking preparation of test pieces, some trend towards the use of directly moulded ones, wherever feasible, was inevitable. That introduces the influence of any melt processing stage into a fatigue study, which can, of course, be the intended purpose, as it was for Crawford and Benham, but otherwise it is a potentially distorting factor. Inevitably, it is necessary to accommodate the processing route and other influential factors in the overall design of fatigue experiments.

If lifetimes are governed by the number of stress alternations and frequency is not influential, the use of high test frequencies could minimize the duration of experiments. Unfortunately, most plastics are prone to a rise in temperature under repeated stress cycles because of their relatively high $\tan \delta$ values in specific regions of frequency and temperature coupled with their low thermal conductivities, see Chapter 5. Thermal failure might then ensue, see Section 10.10; and in less severe cases, with thermal failure not directly manifest, the rise in temperature might, nevertheless, distort the results and any inferences based on them.

Overall, a comprehensive evaluation of fatigue resistance requires quantification of the effect of many influential variables or alternatively their control during the test programme and specification at the reporting stage. Sims and Gladman[3] list nine such factors in their recommendations for fatigue test programmes for glass fibre reinforced plastics.

Crack growth in fatigue is the second approach used for studying fatigue behaviour of plastics and composites. In these types of experiment it is recognized that cyclic loading will probably lead to brittle fracture and therefore the experimental strategy is to measure crack growth in fatigue. A sharply notched test piece is used for this purpose (a compact tension test piece is common) and is deformed in load control with a cyclic waveform:

$$\Delta P = P_{max} - P_{min} \qquad\qquad [S10.7.1]$$

The maximum force (P_{max}) is usually in tension in order to open the crack and in order to avoid any buckling instabilities P_{min} is also often greater than zero tensile force. The parameter R defines this ratio:

$$R = \frac{P_{min}}{P_{max}}$$ [S10.7.2]

Positive values of R relate to tension–tension fatigue tests and negative values relate to tension–compression fatigue tests (other variations can be imposed).

Crack length, a, is either measured during the test using a video camera or a crack gauge or calculated via the measured compliance. The data are then tested in terms of the Paris law equation:[4]

$$\frac{da}{dn} = A(\Delta K)^m$$ [S10.7.3]

where n is the number of fatigue cycles, A and m are constants and the change in the stress field intensity factor (ΔK) is given by:

$$\Delta K = \Delta \sigma_R Y a^{\frac{1}{2}}$$ [S10.7.4]

where Y is the shape factor and σ_R is the stress remote from the notch. Therefore, the calculation of ΔK will depend on knowledge of the dimen-

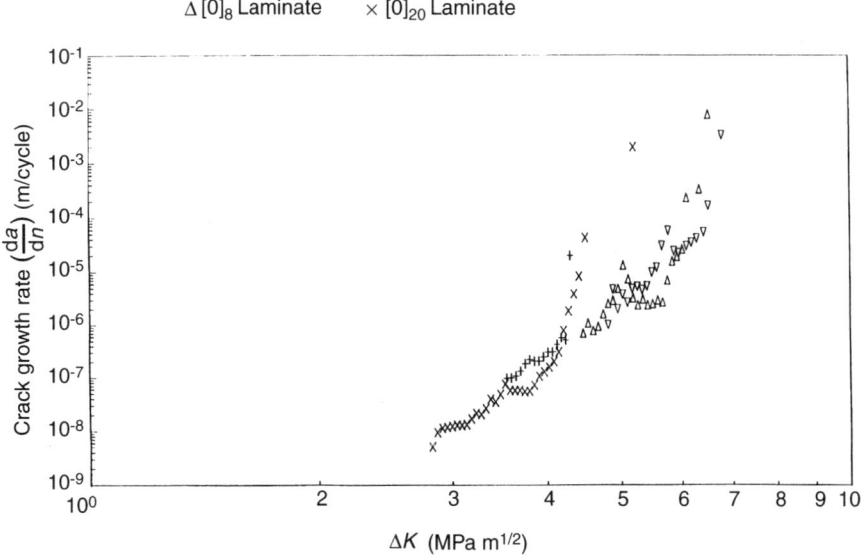

S10.7/1. Fatigue crack growth in unidirectional laminates of AS4/PEEK with 8 and 20 plies.

sions of the test piece, the applied maximum stress and the changing crack length.

The Paris law is a surprisingly regular representation of the data, at least for those conditions away from a threshold of crack initiation and away from gross instability. For example, Fig. S10.7/1 shows some data for crack growth in continuous carbon fibre reinforced poly ether-ether-ketone (AS4/PEEK) composites in the form of unidirectional laminates. The cracks are growing in an intralaminar manner (between the fibres). These particular results, expressed as plots of $\log(da/dn)$ vs. $\log \Delta K$, are consistent with the Paris law integer (m) and intercept (A) being the same for different numbers of plies in the consolidated composite sheets.[5]

The Paris law approach has been widely used for polymers and composites for engineering design and materials comparison.[6–8]

References

1 M. J. Owen. 'A new fatigue testing machine for reinforced plastics.' *Transactions Journal of the Plastics Institute*, 35, 353, 1967.
2 R. J. Crawford and P. P. Benham. 'Some fatigue characteristics of thermoplastics.' *Polymer*, 16, 908, 1975.
3 G. D. Sims and D. G. Gladman. 'A framework for specifying the fatigue performance of Glass fibre reinforced plastics.' NPL Report DMA(A) 59. December 1982.
4 P. C. Paris. PhD dissertation (Lehigh University) Sept. 1962.
5 D. R. Moore and J. C. Seferis. 'Intrinsic characterisation of continuous carbon fibre thermoplastic composites–3 Fatigue crack growth.' *Pure and Applied Chemistry*, 69, 5, 1153, 1997.
6 R. W. Hertzberg and J. A. Manson. *Fatigue of Engineering Plastics*, Academic Press, 1980.
7 D. R. Moore. 'Fatigue of thermoplastics' Ch 10 Vol 7. *Composites Materials Science: Thermoplastic Composite Materials*, ed. L. A. Carlsson, Elsevier, 1991.
8 L. Castellini and M. Rink. Chapter on 'Fatigue crack growth of polymers' in *Fracture Mechanics Testing Methods for Polymers, Adhesives and Composites*, ed. D. R. Moore, A. Pavan and J. G. Williams. ISBN 008 0436897, Elsevier, 2000.

S10.10 The influence of thermal heating on the fatigue behaviour of continuous carbon fibre PEEK composites

The consequence of internal heating that may arise during fatigue testing has been studied for continuous carbon fibre (AS4) reinforced poly ether-ether-ketone (PEEK) composites.[1] 16-ply laminates in two cross-ply configurations were studied:

1 axial test pieces cut from $[-45, 0, +45, 90]_{2s}$ laminates;
2 ±45° test pieces cut from $[0.90]_{4s}$ laminates.

The objectives of the study were to explore the influence of waveform frequency on the fatigue strength in simple zero tension, load controlled experiments for waisted tensile test pieces. Square waves were used at two test frequencies of 0.5 Hz and 5 Hz. Figures S10.10/1 and S10.10/2 show the S–N fatigues curves. For the $[-45, 0, +45, 90]_{2s}$ laminate there is a noticeable reduction of fatigue strength at the higher test frequency and this is more than consolidated by the results for the ±45° test pieces cut from $[0.90]_{4s}$ laminates.

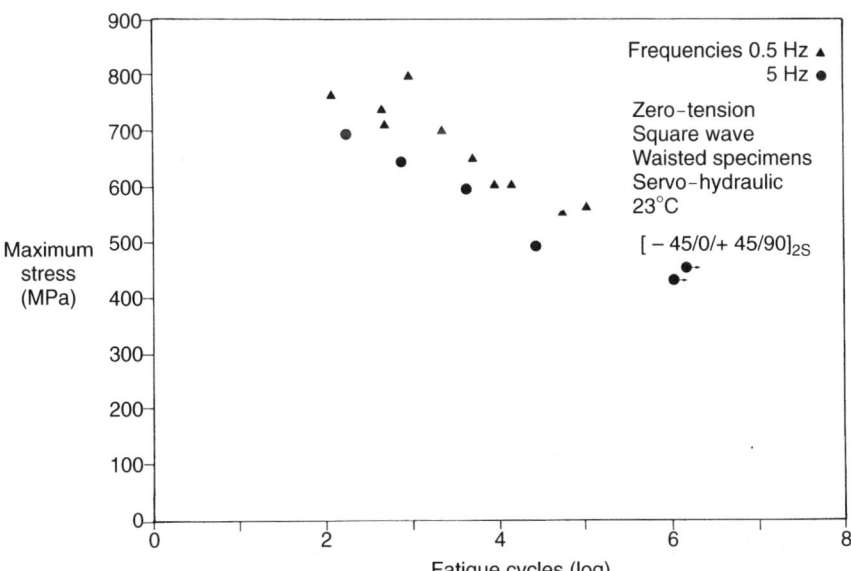

S10.10/1. The influence of test frequency for $[-45, 0, +45, 90]_{2s}$ laminates.

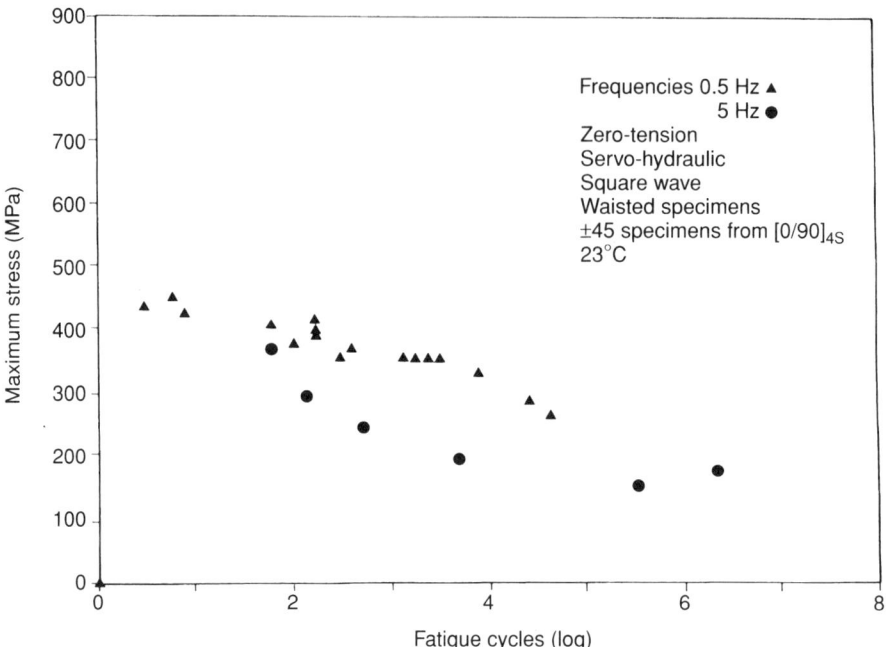

S10.10/2. The influence of test frequency for $[0, 90]_{4s}$ laminates.

During the fatigue experiments, thermocouples were placed on the test pieces, enabling any temperature rise to be monitored during the tests. Figure S10.10/3 shows the results for the quasi-isotropic laminate (where additional measurements for some 32 ply laminates are also included). There is a clear rise in temperature dependent on test frequency where the higher the test frequency, the bigger the rise in temperature. However, some other factors need also to be considered,[1] but a temperature rise of about 30 °C is observed for the $[-45, 0, +45, 90]_{ns}$ laminates (where $n = 2$ and 4).

The experiments on the ±45° test pieces cut from $[0.90]_{4s}$ laminates show even more dramatic temperature rise, as shown in Fig. S10.10/4. Temperature rises of more than 150 °C are now observed and again, the higher the test frequency, the higher the temperature rise. Accounting for these affects is not entirely straightforward.[1] Nevertheless, the test frequency influences the possible rise in temperature of a test piece during fatigue and, in turn, this can influence the outcome of a test.

Although the full ramifications of this are not discussed here, it is clear that such phenomena should be considered in the strategic approach to obtaining strength properties in fatigue. It should limit the use of accelerated testing at high test frequencies unless there is sound evidence of no consequential effects on the outcome.

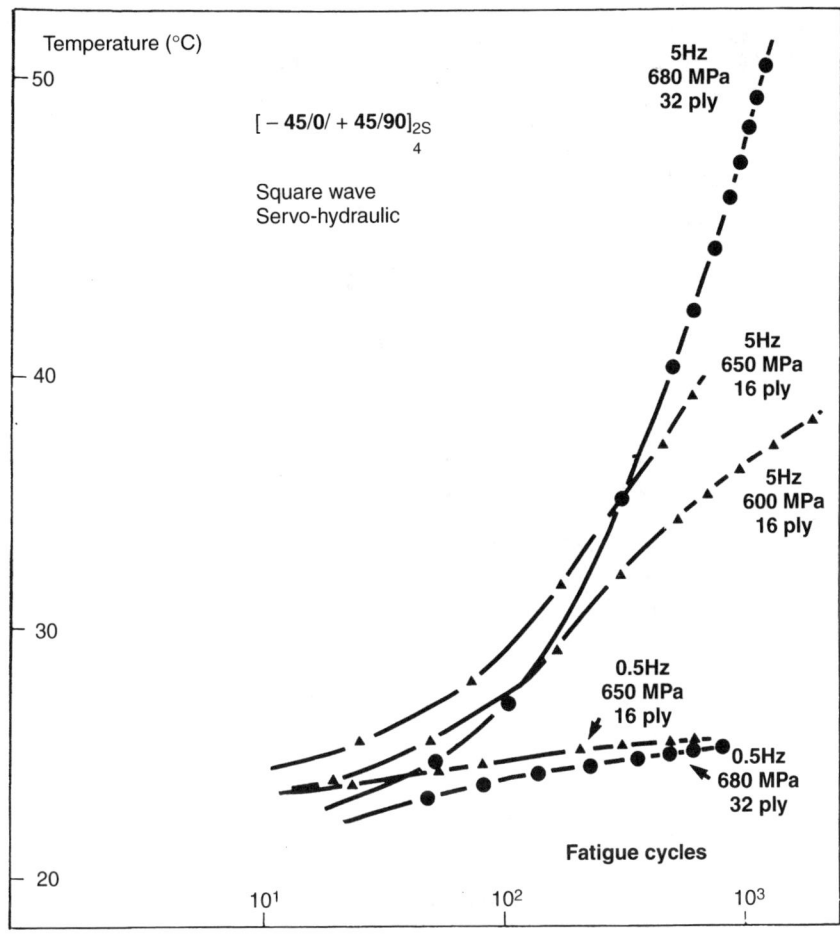

S10.10/3. Specimen temperature rise during fatigue for [−45, 0, +45, 90]$_{ns}$ laminates (where $n = 2$ and 4).

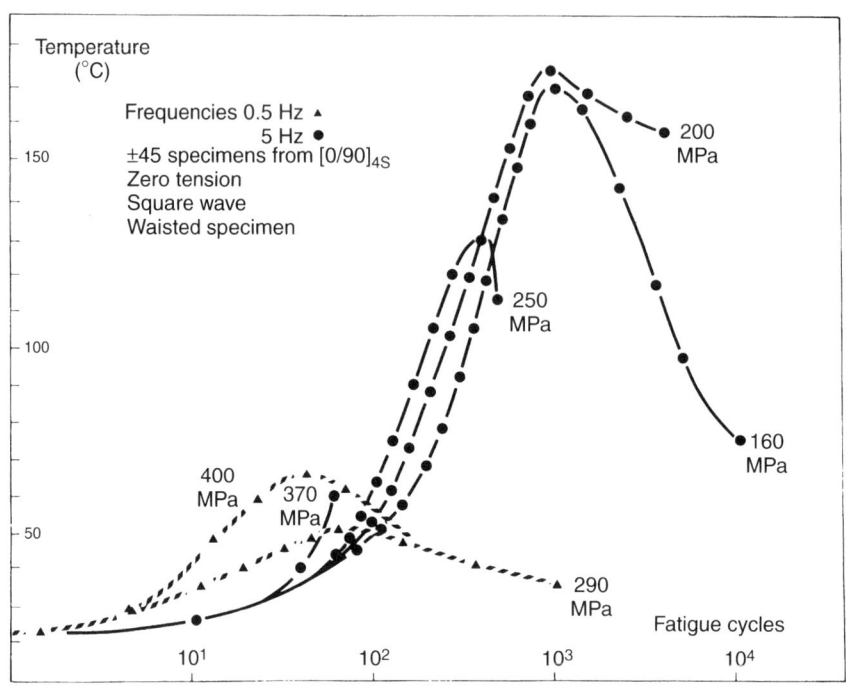

S10.10/4. Specimen temperature rise during fatigue for [0, 90]$_{4s}$ laminates.

Reference

1 D. C. Curtis, D. R. Moore, B. Slater and N. Zahlan. 'Fatigue testing of multi-angle laminates of CF/PEEK.' *Composites*, 19, 6, 446, 1988.

S10.11 Environmental stress cracking

Many years ago Howard proposed the following definition of environmental stress cracking for polyethylene:[1]

Environmental stress-cracking is the failure in surface-initiated brittle fracture of a polyethylene specimen or part under polyaxial stress in contact with a medium in the absence of which fracture does not occur under the same conditions of stress. Combinations of external and/or internal stress may be involved, and the sensitising medium may be gaseous, liquid, semisolid or solid.

A modern definition valid for all plastics requires only small changes.

The polymer industry has seen many cases of interaction between environments (which can be considered as anything from air to a hostile liquid chemical) and plastics (e.g. references 2–6) and this experience provides a useful guide. Two issues have to be addressed in contemplating the interaction. First, does the interaction lead to a primary chemical attack, and second, is there a secondary or physico-chemical attack?[2] For a primary attack, immersion of the plastic in the environment leads to chemical interaction between the environment and the main polymer chain which will become radically changed or broken, i.e. primary chain scission. For a secondary attack, the environment reduces the bonding interactions between the polymer chains. Such secondary attack requires both an active environment and a stress field to trigger the effect, i.e. environmental stress cracking (ESC).

The hostility of an environment depends on its interaction with the main polymer chains and can be of a chemical or physical nature. Aspects of this are not easy to predict, but the nature of the environment and its potential as an environmental stress cracking agent can be described (e.g. reference 3) by at least two attributes. First, in terms of the size of the molecule, a relatively small molar volume can lead to the substance acting as a hostile environment; thus, for example, methyl etheyl ketone, MEK, which has a molar volume of 90 is a relatively small molecule where a range between 50 and 400 can be anticipated for chemical environments. Second, in terms of the hydrogen bonding parameter for the environmental fluid, a modest value of that quantity can lead to it acting as a hostile agent. For example, MEK has a hydrogen bonding parameter of $5.1\,\mathrm{MPa}^{1/2}$ where a range of 4 to $20\,\mathrm{MPa}^{1/2}$ can be anticipated for most chemical environments, i.e. MEK has a relatively modest hydrogen bonding parameter and on this count also can be expected to be a hostile agent. Therefore, almost without reference to the plastic or composite that may be used with MEK, we can expect MEK to behave in a relatively aggressive manner as an environmental stress cracking agent, though plastics will differ in their resistance to it because of their individual physico-chemical properties. The complex interactions between plastics and environments are not understood at the level of fine discrimination.

Full investigation of secondary interactions requires a combination of stress and the environment because the ability of the plastic or composite to absorb the environment in the absence of a stress field is likely to be small. The presence of the stress field accelerates the absorption of the environment although this can be anticipated to be a process of selective absorption,[2] particularly in the vicinity of flaws in the solid material. In this context, it might be difficult to identify a flaw in a strictly physical sense but the adoption of Griffiths' concept of a flaw can readily accommodate these arguments.

The experimental techniques for studying ESC are essentially those employed in creep rupture studies, i.e. at nominally constant stress, or at nominally constant strain, but in the presence of active environments. The deformation mode is usually either tension or flexure. The former was adopted by Lander[7] for tests on the polyethylenes and plastics of similar modulus and has been widely used, with elaborations, for many classes of plastic ever since. Flexure is used for many of the lower-modulus plastics, partly for convenience and partly because the critical strain for the onset of crazing or other visible signs of attack can often be determined in three-point bending where the strain at the surface varies along the beam. The Bell Telephone Laboratories (BTL) Test, first proposed by De Coste, Malm and Wallder[8] and later formalised as ASTM D1693-66 (*Test for Environmental Stress Cracking of Type 1 Ethylene Plastics*) also imposes flexure but the deformation is very severe, deliberately so because of the nature of the physical problem and the commercial issues that it was originally developed to resolve. The step excitation of these tests is sometimes replaced by a ramp.

Brown[9] published a critical survey of various test methods including an assessment of the relative merits of constant stress (which he preferred for design purposes) and constant strain experiments (which he thought had value for quality control). Despite much experimentation since that survey was written there has not been any radically new developments in the approach, presumably because Howard and the other early workers in the field had correctly identified the critical test variables and the concept of mechanical testing in the presence of an aggressive environment is quite basic.

A comprehensive investigation requires that the response to different environments be studied over a range of temperatures and excitation amplitudes but this is rarely practicable and the standard recommendations limit the ranges, but specify levels likely to ensure failure within a conveniently short time. These techniques of accelerated testing are acceptable in so far as the Larson–Miller parameter or the time–temperature superposition principle may be applicable, but accelerated stress cracking experiments appear to be susceptible to rather specific irregularities which imply a qualitative rather than a quantitative significance for the data. Nevertheless, the Lander test has the attributes of a genuine physical test and is complicated only in that the results are sensitive to various largely unquantified interactions between the specimen surface and the environment. The BTL test, on the other hand, is a hybrid. It was introduced to simulate the possible stress state in the insulation of a communications cable subjected to bending and to induce environmental stress cracking in what is an exceedingly tough class of polymer. It subsequently acquired an overriding status as the quality control test for polyethylenes used as insulation in underwater cables or sheathing and as the database grew and commercial positions were consolidated any inclination for change to a more scientifically definitive test became of secondary importance. The test suffers from the general deficiency of all hybrid tests that all aspects have to be controlled closely and arbitrarily if results are to be reproducible and the constraints entailed by this requirement coupled with those necessary to ensure substantial acceleration of the failure processes resulted in a test procedure that appears to be largely unrelated to the service conditions for which a performance rating is sought, i.e. it seems highly unlikely that there can be anything more than a correlation between the crack resistance of a highly annealed, compression moulded, severely notched polyethylene beam, highly strained and immersed in an aggressive environment at 50 °C, and the service performance of an extruded cable insulation that may not even be in contact with its environment of sea water mainly at 4 °C. However, a superior alternative never appeared and there is ample evidence that the arbitrary BTL method has satisfactorily served its intended purpose. The nominally superior Lander test, on the other hand, fails to discriminate between the tougher grades of polyethylene, which can be ranked by the BTL test. The former is not severe enough, even when notched test pieces are used; the BTL test, on the other hand, applies a severe combined stress field to the test piece by a very simple practical expedient.

Initially, the use of standardized notched specimens was a practical expedient rather than a deliberate adoption of the principles of fracture mechanics. Gotham briefly discussed the merits of notched specimens[10] in relation to results for an injection moulding grade of polyethylene (which was much more susceptible to environmental attack than the cable grades which would be immune to his test conditions). Cracks initiate at surface imperfections and edge defects[11] but such effects are overwhelmed if the specimen is severely notched and experimental scatter is reduced. Marshall et al.[12] considered the notch tip geometry and interpreted their results in terms of linear elastic fracture mechanics. Their specimens were 150 mm long, 50 mm wide and of three different thicknesses, approximately 15 mm, 31 mm and 46 mm. They plotted crack length vs. the associated K_c value for various elapsed times and found that the results from many separate experiments with different initial crack lengths fell on a common curve within a scatter band of ±12%. The scatter along the log (crack speed) axis was wider than a decade, similar to the normal range of lifetimes observed in tests on unnotched specimens and therefore their results may have been dominated by inherent flaws rather than by the notch. Numerous investigations of the same general type followed.

An illustration of these combined considerations has been documented for the environmental creep rupture behaviour of polyimide films.[13] Two such films are used in electrical applications where load-bearing resistance in hot wet environments is critical. Two contending polyimde films were prepared with the same aromatic diamine but different aromatic dianhydrides, namely diamino diphenyl ether (DDE) with either pyromellitic dianhydride (PMDA) or with 3,4,3′,4′ biphenyl tetracarboxylic dianhydride (BPDA).

The performance of both films in tensile tests after conditioning in water is identical. When these 50 μm films were tested under load in a demineralized water environment, a different view of their relative performance emerged, as in shown in Figures S10.11/1 and S10.11/2. The results for the BPDA/DDE polyimide are shown in Fig. S10.11/1 for tests at 70 °C and 90 °C in air and demineralized water. All test pieces fractured in a ductile manner and although there is some scatter in the data it could be concluded that a hot/wet environment (e.g. 90 °C in water) did not result in a reduction in tensile strength. Figure S10.11/2 shows the results for the other polyimide, PMDA/DDE, where similar strength is observed for the tests in

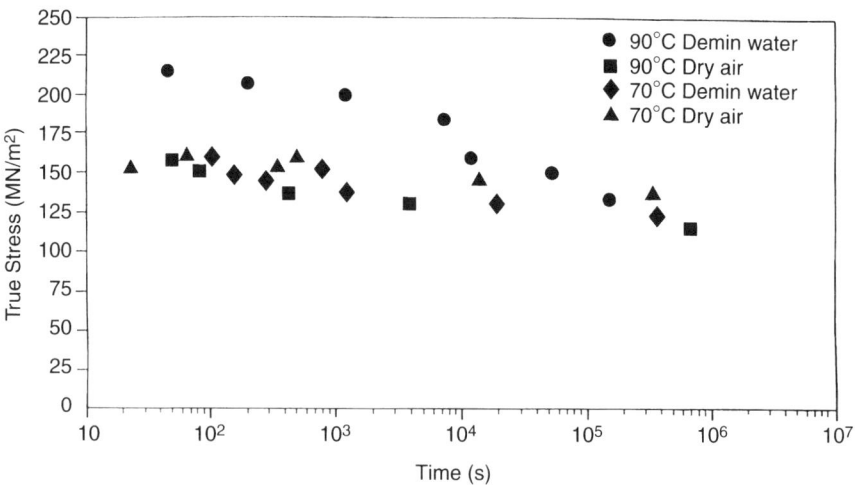

S10.11/1. Environmental creep rupture for BPDA/DDE polyimide film.

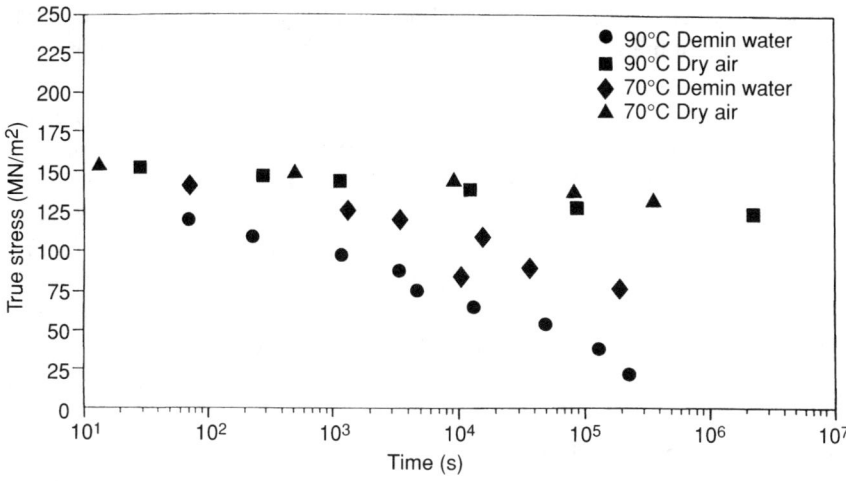

S10.11/2. Environmental creep rupture for PMDA/DDE polyimide film.

air, but reduction in strength is observed for the hot/wet environments. Moreover, a transition in fracture from ductile to brittle accompanied the reduction in strength.

In general, the more rigid thermoplastics, especially the amorphous ones, are less tough and crack at much smaller strains than the polyethylenes and polymer film samples. Therefore, quite simple apparatus, for instance an end-loaded cantilever, suffices to generate a stress or strain that varies along the major axis and reveals a boundary beyond which cracking or crazing develops. That boundary, though not necessarily very sharp, defines a threshold strain (or stress) for crazing which is at least semi-quantitative, whereas the BTL test merely gives the failure probability or the distribution of failure times for an arbitrary compound strain.

The observed position of the boundary and the threshold time for the onset of crazing depend to some degree on the intensity and style of the illumination and the visual acuity of the observer or the sensitivity of the optical system. That became a difficulty in the BTL test as increasingly tough polyethylenes were developed for the cable insulation applications, to the point where the identification of threshold time became subjective. Furthermore, the results from laboratory tests on standard specimens often bear no relation to the actual service performance of injection mouldings because of differences in the flow geometries and processing conditions which generate anisotropy and internal strains. That poor correlation between many laboratory tests and service experience can be partly recti-fied by the use of test pieces cut from service items or even tests on the entire item, e.g. immersion in an appropriate aggressive liquid for a short period, see Section 10.12. Criteria of acceptability are established by trial and error, combined with 'calibration' with respect to service experience; for instance, a 'good quality' moulding of a polyethersulphone will not be harmed by immersion in ethyl acetate, whereas it may crack if it is immersed in toluene and will certainly crack if it is immersed in acetone. When such mouldings are immersed in ethyl acetate for several hours a pattern of fine crack-like lines appear. The cracks are aligned with the molecular orientation of the surface layers, see the two photographs reproduced in Fig. 10.12/1 in the main text (but not a polyethersulphone moulding). A 'poor quality' moulding behaves somewhat less robustly and there is also an interacting sensitivity to the molecular weight of the polymer.

During the many intervening years since Howard's pioneering work on polyethylenes, there have been numerous studies of the micromechanics, macromechanics and physico-chemistry of environmental stress cracking but there have been no major advances in the experimental techniques, apart from the application of fracture mechanics. There are extensive databases on the durability of many plastic/environment combinations but industrial need to check any new potentially unsatisfactory combinations entails a lengthy evaluation programme using what are still essentially the methods of the Howard era supplemented only by a more advanced understanding of the effect of sharp notches and an associated wariness over long-term predictions. Reference 14 is a modern review of current test practices for the assessment of resistance to environmental stress cracking.

References

1 J. B. Howard. 'A review of stress-cracking in polyethylene.' *SPE Journal*, 15, 397, 1959.

2 R. P. Kambour. 'Mechanistic aspects of crazing and cracking of polymers in aggressive environments.' GE CRD Report No. 77CRD169, August 1977.

3 R. P. Kambour and C. L. Gruner. 'Effects of polar group incorporation on the crazing behaviour of glassy polymers: styrene/acrylonitrile copolymer and a dicyano bisphenol polycarbonate.' GE CRD Report No. 77CRD206, Oct. 1977.

4 G. A. Bernier and R. P. Kambour. 'The role of organic agents in the stress crazing and cracking of poly(2,6-dimethyl-1,4 phenylene) *Macromolecules*, 1, 5, 393, 1968.

5 D. C. Wright. 'A review of factors that influence EAC.' A RAPRA Technology Ltd. Publication, 1996.

6 L. C. E. Struik. 'Physical aging in amorphous polymers and other materials.' TNO Communication 565, Proefschrift Delft, 1977.

7 L. L. Lander. 'Environmental stress rupture of polyethylene.' *S.P.E. Journal*, 16, 1329, 1960.

8 J. B. De Coste, F. S. Malm and V. T. Wallder. 'Cracking of stressed polyethylene. Effect of chemical environment.' *Industrial Engineering and Chemistry*, 43, 117, 1951.

9 R. P. Brown. 'Testing plastics for resistance to environmental stress cracking.' *Polymer Testing*, 1, 4, 267, 1980.

10 K. V. Gotham. 'A formalized experimental approach to the fatigue of thermoplastics.' *Plastics and Polymers*, 37, 309, 1969.

11 A. O'Connor and S. Turner. 'Environmental stress cracking of an injection moulding grade of polyethylene. Part 1.' *British Plastics*, 35, 452, 1962.

12 G. P. Marshall, L. E. Culver and J. G. Williams. 'Environmental stress crack growth in low density polyethylene.' *Plastics and Polymers*, 38, 95, 1970.

13 D. R. Moore, M. S. Sefton and J. M. Smith. 'The environmental creep rupture behaviour of polyimide films produced using pyromellitic dianhydride (PMDA) and 3,4,3',4' biphenyl tetracarboxylic dianhyride (BPDA) precursors.' *Plastics Rubber and Composites: Processing and Applications*, 18, 281, 1992.

14 A. Turnbull and T. Maxwell. 'Test methods for environmental stress cracking of polymeric materials.' *N.P.L. Technical Review*, No. 3.

S10.15   The engineering design of water storage tanks from creep rupture and fatigue data

The way in which a vertical cylindrical tank distorts when filled with a liquid was determined by Timoshenko and Woinowsky-Krieger[1] but simplified by Forbes *et al.*[2] when they applied it to the design of large liquid storage tanks fabricated from extruded sheet. With a modest degree of over-design the engineering equation reduces even further to:

$$\frac{\sigma_D}{\rho} > \frac{gHr}{d} \qquad\qquad [S10.15.1]$$

where σ_D is a design stress, ρ is the density of the liquid to be stored in the tank, d is the tank wall thickness, H the height of the liquid level and r the tank radius.

The equation was used by Moore and Gotham[3] in the design of rotationally cast polyethylene tanks for cold water storage. The engineering problem was to define a magnitude for design stress for the polyethylene and hence determine the appropriate wall thickness for the tank of known height and radius in order to avoid the tank breaking during a required 5-year lifetime. The materials science aspects of the problem were tackled by a combination of creep, creep rupture and fatigue experiments on rotationally cast polyethylene samples. Figure S10.15/1 shows tensile data for

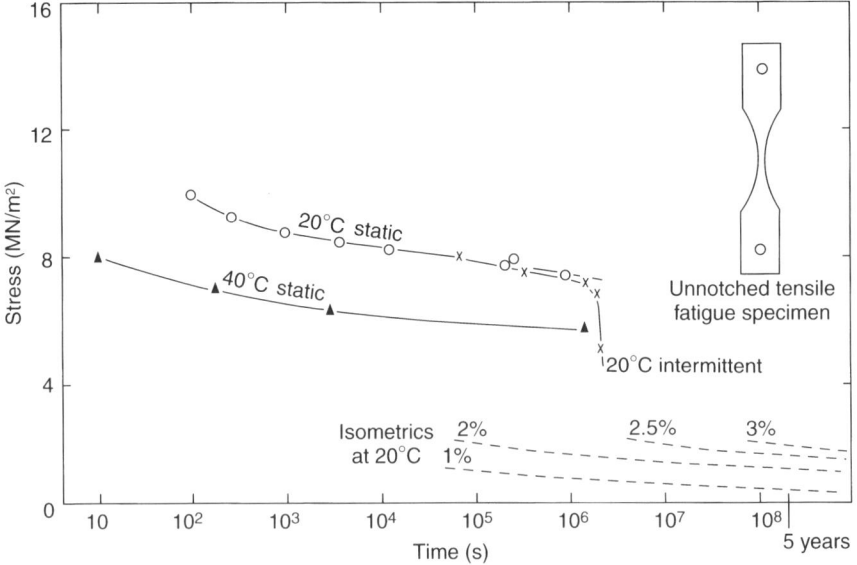

S10.15/1. Fatigue, creep rupture and isometric creep data for rotationally moulded polyethylene at 20 °C and 40 °C.

these experiments for unnotched specimens (illustrated in the figure). The creep data have been added to the stress vs. log time under load plot in the form of isometric data. The creep rupture experiments all indicate ductile fracture in the polyetheylene samples, but cyclic loading was shown to introduce embrittlement after a few weeks' testing. The cyclic loading involved 8 hours on load and 16 hours off load and the data are plotted as total time under load. The prospect of crack initiation in service could reduce the lifetime of the tank to less than 5 years and therefore additional fatigue experiments, involving notched specimens, gave an indication of the level of brittle strength. These data are shown in Fig. S10.15/2. They show that a blunt notch (tip radius 250 μm) does not precipitate severe embrittlement but that as the notch severity is increased (to a sharp crack of tip radius 10 μm) then a ductile–brittle transition in fracture may occur. The brittle strength value was commensurate with a strain level of about 0.025, which had earlier and coincidentally been successfully adopted, in the light of extensive downstream experience, as a limiting value for safe load-bearing design for many grades of polyethylene. Therefore, that value was used in this particular case to obtain the 5-year design stress after some derating for the influence of temperature in service.[3]

A design stress of 1.13 MPa emerged from these studies and this was applied to the engineering design formula of Eq. S10.15.1. A number of

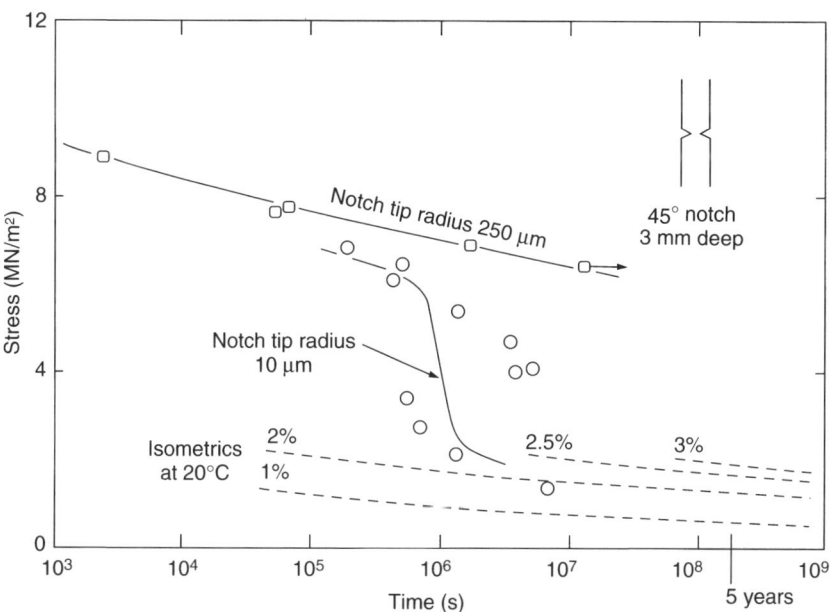

S10.15/2. Fatigue on notched rotationally moulded polyethylene at 20 °C (8 hours on load: 16 hours off load).

tank wall thicknesses were then established for a wide range of tank sizes. For liquids other than water, the wall thicknesses would have been different according to the density of the stored liquid but additionally the possibility of environmental stress cracking would have been investigated. In the case of liquids only mildly aggressive towards polyethylenes and otherwise harmless, a lower limiting value of the strain, and hence of the design stress, would have been adopted. In the case of the containment of a liquid severely aggressive towards polyethylenes, a different polymer system would have been sought, using the existing databases and the current suite of environmental stress cracking tests to establish a safe limiting strain. In the case of hazardous liquids, other considerations would have prevailed.

References

1 S. Timoshenko and S. Woinowsky-Krieger. *Theory of Plates and Shells*. McGraw Hill, 1959.
2 K. Forbes, A. McGregor and S. Turner. 'Design of fluid storage tanks from polypropylene.' *British Chemical Engineering*, Oct. 1970.
3 D. R. Moore and K. V. Gotham. 'The design of cylindrical tanks in rotationally cast polyethylene from strength and deformation studies.' *Plastics and Polymers*, p. 30, 1975.

11

Ductility assessment from impulse excitations

11.1

An impulse excitation is conveniently tractable within the theory of linear viscoelasticity and is readily linked to the ramp, step and sinusoidal excitations via the linear superposition principle. It is also conveniently practical, as an impact test, even for non-linear viscoleastic systems. In an impact test the impulse is one of energy and the response is work done on the impacted item, i.e. the development of both a force and a displacement.

When the incident energy is much higher than that necessary to induce failure (excess energy test), the initial response of an impacted test specimen is similar in appearance to the response to a high rate ramp excitation. However, as energy transfers from the impactor to the specimen, the straining rate decreases progressively and any similarity to a ramp function disappears. When the incident energy is lower than that needed to induce failure, the impactor may be arrested and may rebound from the surface of the specimen. Low incident energy tests are now used widely and give useful information about the early stages of failure, but the normal and mainly standardized procedures employ incident energy in excess of that necessary to induce failure.

11.2

In the plastics industry, impact tests are used primarily for an assessment of toughness under conditions vaguely similar to those that prevail during many impact events in service. However, many practical impacts entail much higher velocities of impact and, through the additional agency of geometrical features and constraints, higher straining rates. There is little that can be done to remedy this situation either by faster impact tests or by conventional ramp excitation machines. On the one hand, the generation of high ramp excitation rates is difficult and certainly expensive to engineer. On the other hand, instrumented tests, in which force–time signals are monitored, are already at the limit of their capability since transducer frequency response interacts with stress wave frequency in the test piece. Impact tests fall mainly into three groups, as defined by the deformation mode, namely tensile, flexed beam and flexed plate. There are two versions of the flexed beam method; in the Izod test the beam is a cantilever and in the Charpy test it is a freely supported beam impacted at the centre. There are also two principal versions of the flexed plate method; in one the specimen is freely supported and in the other it is clamped at the support, which in both cases is an annulus if the test material is a thermoplastic, but may be square or rectangular if it is a fibre composite. Another variant has the striking face of the impactor flat rather than hemispherical, for which certain advantages are claimed.[1]

These widely-used test procedures give superficially similar data, but they tend to measure different attributes of the impacted item and consequently the data differ quantitatively. The phenomena are difficult to interpret in terms of the conventional physical properties for several reasons, viz.

1   The stress field at the site of impact is complex and changes rapidly as the test piece deforms and/or damage develops.
2   The pattern of shock waves generated by the impact is a function of the properties of the test piece, its geometry and the geometry of the apparatus.
3   The test piece is itself a structure, with properties that commonly vary with direction and from point to point.

Because of those interacting factors, impact testing developed somewhat arbitrarily, with less regard for scientific principles than for ease of operation and supposed simulation of impacts in service. The latter objective remains elusive; despite supportive anecdotal evidence, there is no simple link between the impact resistance of a laboratory test specimen and that of an end-product made from the same material, though there are useful correlations. Furthermore, each aspect of impact resistance reflects several physical properties, to the point that they are best regarded as attributes rather than simply definable properties.

11.3

Various national and international standards specify the details of the test methods, which are widely used for the generation of data to support the businesses of the primary producers of materials, the converters and the users of plastics end-products. The different classes of test are often used in combination but certain patterns of usage that have emerged reflect regional and sector preferences. Thus, in the plastics industry, data utilizers in the USA usually require Izod test data (ASTM Method) and those in Continental Europe usually require Charpy test data (DIN or ISO Method); in both areas some require flexed plate data and a few require tensile impact data. In the composites industry the Charpy and the flexed plate configurations are the common choices but the latter often differ from the standard configuration used by the thermoplastics industry in that the test pieces can be relatively large square or rectangular plates, probably as simulations of structural panels.

11.4

The energy impulse is usually delived via a swinging pendulum for the flexed beam tests and via a falling mass, often referred to as a 'dart', for the flexed plate tests. However, the modes of delivery are sometimes interchanged. Pneumatically driven projectiles and hydraulically driven rams are sometimes used to achieve higher, but slightly unpredictable velocities of impact, usually without radical change to the test configurations. Another method, which does entail a different configuration, delivers the energy impulse via a shock tube.[2] That method is for the purist; it delivers a well-defined impulse but it entails relatively elaborate apparatus and is not versatile in regard to the test specimens.

In all the classes of impact test there are options for notched specimens from tests on which the deleterious effects of stress concentrators can be measured or inferred. From time to time, also, investigators modify a standard test configuration or procedure in order to pursue a particular issue. Thus, for instance, Thomas and Meyer[3] modified an Izod impact tester so that very small beams cut at various points from a standard injection-moulded tensile bar could be impacted in a miniature version of the cantilever beam configuration. In view of certain practical difficulties over the clamping of cantilevers the accuracy and precision of the results by Thomas and Meyer were probably low, though the modified apparatus served their purpose well. Similar miniaturizations of the Charpy and flexed plate tests are not feasible but both can be modified to accept larger test pieces than the standard ones.

The flexed plate configuration is often changed to a 'flexed item' configuration, the item being variously plates with surface features, boxes, bottles, other three-dimensional shapes (which may be actual service items) and extruded pipes. Such shapes should be gripped or constrained in specific ways to ensure that the energy absorbed by elastic and plastic distortions of the entire structure are identifiable and fairly constant from test to test. Figure 11.4/1 shows a typical modification by which the relative impact resistance and/or penetration resistance of sections of pipe can be measured. Meijering discussed the impact testing of sections and segments of pipes[4] and Terselius et al.[5] modified the plate support to accommodate

11.4/1. Arrangement for impact tests on sections of extruded pipe by the falling dart method.

curved plates cut from pipe in their studies of the effect of gelation on the impact resistance of PVC pipes.

No quantitative correlations have been established between the data derived from the different tests, except possibly between Izod and Charpy data when the fracture is brittle.[6] That lack of quantitative correlations arises in part from the inherent complexity of each impact event, set out in Section 11.3, and additionally from the following factors.

1   Each test configuration interacts in a unique way with the flow geometry in the critical region of the test piece, and many of the earlier attempts to establish correlations were flawed in that they failed to use a common flow geometry for the paired tests.
2   Poor experimentation, due largely to misconceptions regarding attainable precision that have arisen from an earlier collective failure on the part of investigators to identify the main sources of error.

The mechanical origins of experimental error in impact testing are discussed in Supplement S11.4.

11.5

Originally, impact testing was a relatively crude operation. A known and overwhelming energy was applied and the residual energy in the impactor after impact was measured, from which the energy absorbed during the impact event could be deduced. The absorbed energy was referred to as the 'impact energy', even though some of the energy is absorbed by associated mechanisms such as friction in the test machine, kinetic energy in the specimen and acoustic energy.

In a variant of the excess energy method, the incident energy just sufficient to cause failure is derived by a cumbersome but simple procedure in which the severity of each test is chosen on the basis of the outcome of the previous test, e.g. the so-called 'staircase' method or the 'Probit' method. That procedure is often impaired by uncertainty as to how failure should be defined, although it might actually be a more realistic measure of the 'impact energy' than the excess energy method and can still play a useful role in evaluation programmes. The impact energy, as derived by the staircase method, being just sufficient to induce 'failure', is usually less than that measured by the excess energy method. Thus, for instance, it was so in 11 out of 12 comparisons carried out on various glass fibre reinforced nylons, the greatest disparity being the excess energy method recording an energy 1.59 times that given by the staircase method.

Prior to the development of instrumented impact tests, the usual practice was to apply the rapid and simple impact test procedures to relatively large numbers of nominally identical test pieces and analyse the results statistically to counter the prevailing experimental uncertainties and variabilities. The staircase method similarly entails the use of many test pieces. Thereafter, innovations in apparatus, particularly in relation to sensors and data processing, have enabled any test event to be scrutinized in great detail and have changed the emphasis of impact testing to some degree; with deeper analysis of results from fewer tests and an element of diagnosis. Even so, the interpretation of results is still not unambiguously quantitative, mainly because of the transient nature of the successive stages in the impact event but also because the observed response is that of the entire mechanical and electronic system. Even if the electronic part has adequately fast data capture characteristics, the first part of the observed response to the energy impulse may be either delayed and damped by the inherent viscoelasticity of the test piece or contaminated by extraneous vibrations if viscoelastic characteristics are suppressed (i.e. if the material is in its brittle or near-brittle state).

Nevertheless, the now commonly available record of the force–time relationship, and preferably that of the force–deformation relationship, of the impact event enables the analysis of performance to be more diagnostic, more comprehensive and more quantitative than was possible in the era of uninstrumented impact testing. In particular, within the limitations imposed by extraneous noise, the absorbed energy and the deformation associated with any point on the response curve can be computed (see Supplement S11.5), though it transpired that clearly identifiable features on a response curve are not necessarily indicators of specific stages in the development of impact damage, see Section 11.10.

It is salutary to reflect, however, that the main characteristics of impact resistance and the factors affecting it were established by the cruder, uninstrumented tests. The most comprehensive mapping so far of the behaviour of plastics materials under impact was by Vincent, whose monograph[7] remains as valid today as when it was published in 1971.

11.6

Since a derived force–deformation relationship is the overall response of a test piece, which does not necessarily reflect the local situation at the failure site, the computed energy is the total energy imparted to the test piece rather than the energy density at some critical point such as the point at which fracture or yielding initiates. Additionally, failure does not necessarily initiate at or near the point of impact; it occurs where and when the local stress exceeds the local strength. Figure 11.6/1(a) shows a very brittle fracture under flexed plate impact of an injection-moulded disc of polyethersulphone that initiated at an inclusion situated by chance near the support ring, where the local stress should have been very much lower than that at, or near, the point of impact, where the force was measured, but where it had been raised by the stress concentrator. There are two points worth noting. The first is that the highly-stressed central region shows evidence of some ductility, with which the associated force–time record, Fig. 11.6/1(b), is consistent. The second point is that the extremely brittle character of the fracture that terminated the test may have been attribut-

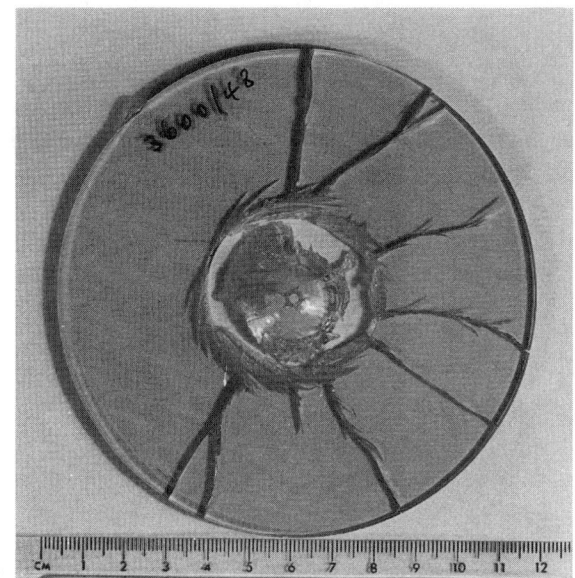

(a)

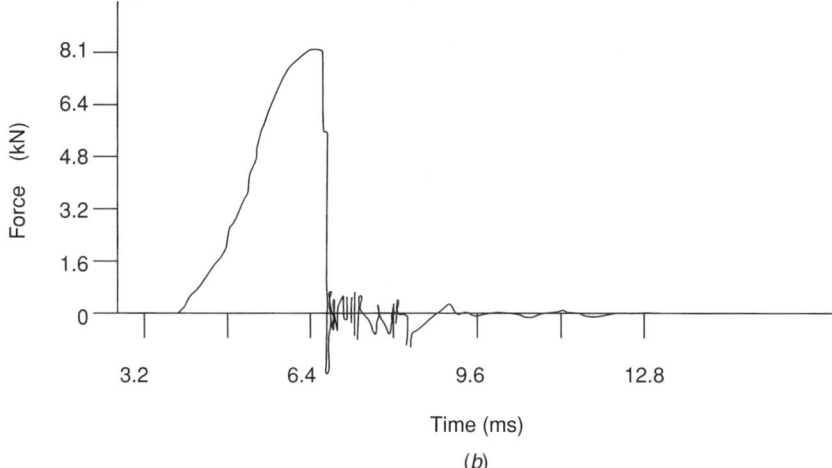

(b)

11.6/1. Brittle failure originating away from point of impact. Injection moulding of poly(ether sulfone): (*a*) broken specimen; (*b*) impact response curve, the shape showing some ductility (at the point of impact).

able to a triaxial stress at the included particle which was likely, because of its composition, to have absorbed some water and to have expanded relative to the surrounding polymer. It is interesting, also, that the path followed by the main crack was strongly influenced by the support ring. That example demonstrates that the appearance of the tests piece should always be taken

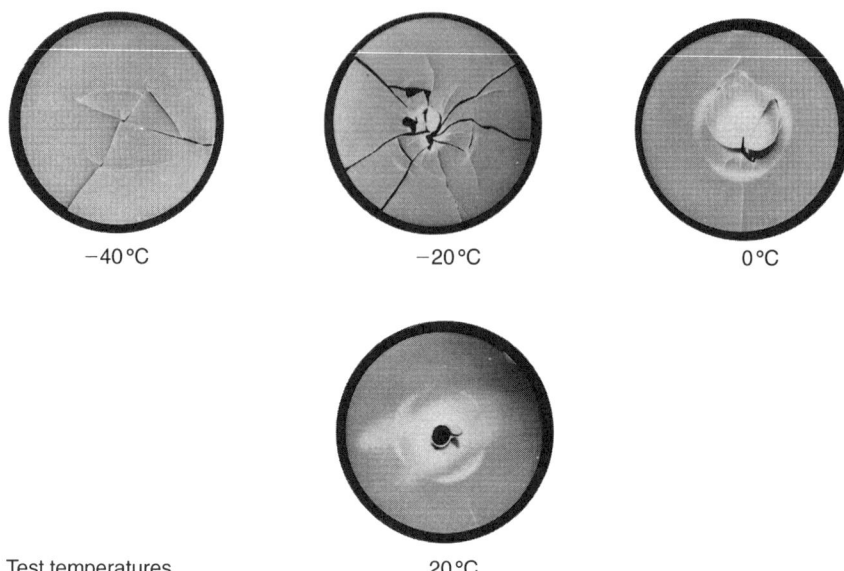

Test temperatures

11.6/2. Failure under impact of discs injection-moulded from propylene–ethylene copolymer. Tests at temperatures in the brittle–ductile transition region produce mixed-mode failures.

into account when impact test data, and strength test data in general, are being assessed.

Even when failure occurs at, or near, the point of impact a quantitative interpretation of the event may not be straightforward because there may be elements of both brittleness and ductility, ranging from brittle fractures with crack tip ductility to extensive plasticity with crack-like tears. Mixed-mode failure can be seen in some of the photographs in Fig. 11.6/2. The damaged test pieces were edge-gated injection-moulded discs of a propylene–ethylene copolymer which had been subjected to flexed plate impact at different temperatures. The top left-hand photograph shows brittle failure that occurred at –40°C and the bottom one shows ductile failure, in this case puncture, that occurred at 20°C. The white ring in the latter is phase separation in the moulding caused by reaction along the line of the supporting annulus when the impactor struck the disc. The other photographs show combinations of brittle cracking and ductility under impact at the intermediate temperatures –20°C and 0°C.

11.7

Since impact resistance, in common with most mechanical properties, is a function of temperature, the best testing practice requires data to be acquired at various temperatures. However, practical difficulties, arising largely from the nature of impact test procedures, result in the achieved temperatures (other than 'room temperature') usually differing significantly from the target temperature. The traditional method was to transfer test pieces rapidly from a temperature-controlled storage chamber to the test machine and impact them immediately thereafter, but calculations and measurements show that a test piece initially at, say, $-20\,°C$ would be $5\,°C$ warmer at its surface within a few seconds of exposure to a room temperature environment. Merle et al.[8] described a device that reduces, but does not completely eliminate, the disparity. A simple solution to temperature control below room temperature of compact test machines such as pendulum actuated Charpy and Izod machines is to follow Vincent[7] and operate them in freezer cabinets. With other methods, the transition temperature can seldom be determined accurately.

The relationship that emerges from tests at various temperatures is one of impact energy (say) at a low value commensurate with the brittle state initially rising very slowly with increasing temperature and then much more rapidly as the brittle–ductile transition region is reached. As the temperature increases further, the energy may decrease, in association with the decreasing yield stress, and as the specimen becomes increasingly prone to plastic instability. Figure 11.7/1 shows the overall relationship schematically; it is analogous to Fig. 9.2/2 in which tensile strength is shown schematically as a function of temperature. The situation in the transition region can be confused because of the mixed-mode failures mentioned previously and associated bimodal distributions of absorbed energy. In the case of the tests from which Fig. 11.6/2 was derived. the distributions of failure energies at $-20\,°C$ and at $0\,°C$, i.e. two temperatures in the transition region, were bimodal and the notional coefficients of variation were correspondingly much higher than those for the tests at $-40\,°C$ and $20\,°C$, where the failures were clearly brittle and ductile, respectively.

The brittle–ductile transition temperature, if Fig. 11.7/1 can be so simplified, is sensitive to the same influential variables that affect the mechanical properties generally, including the straining rate, the specimen size, the stress geometry and the state of molecular order in the test specimen. However, except in a few special cases, the brittle–ductile transition is not closely associated with the glass–rubber transition, which tends to be the dominating factor for the modulus-related properties and many others.

The rate sensitivity is a fundamental feature of viscoelasticity. The effect of specimen size arises because small items are less prone to brittle failure than large ones. The effect of stress geometry arises via the degree of stress or strain triaxiality which affects the molecular relaxation processes (the bulk relaxation processes, i.e. those associated with a triaxial stress field, are slower and more limited in scope than the shear relaxation processes). Some

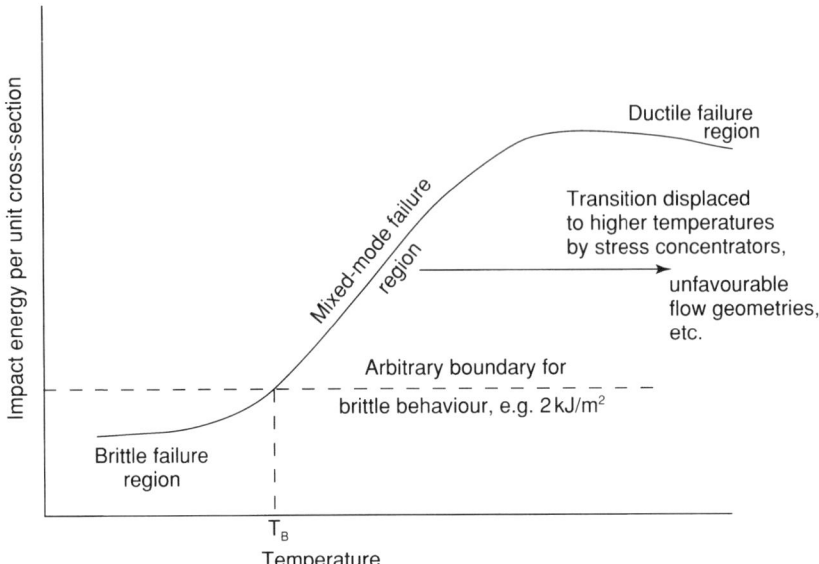

11.7/1. The brittle–ductile transition as a function of temperature. The transition is rarely abrupt; the 'transition temperature' is defined arbitrarily and is sensitive to many factors.

of these considerations are further exemplifed in Supplement S11.7, where a coating of a thermosetting resin on a substrate exhibits a ductile–brittle transition. Moreover, by adoption of linear elastic fracture mechanics, it can be shown that the location of the ductile–brittle fracture transition can be predicted in terms of coating thickness and impact temperature.

11.8

The potentially embrittling effect of stress triaxiality is crucial to all considerations of end-product design, since various geometrical features introduced for aesthetic, functional and economic reasons often introduce stress triaxiality if the item is stressed. The embrittlement is worsened further if the influential feature also entails a detrimental flow geometry and therefore notched specimens are often used in impact testing programmes. Machined notches give information primarily on stress-field embrittlement, though the results are inevitably sensitive also to the molecular state in the vicinity of the tip of the notch; moulded notches give similar information but the results often differ from those relating to machined notches because the straightforward stress concentration effect may be mitigated or exaggerated by local anisotropy arising from the flow geometry and possibly by flow irregularities in the vicinity of the notch. It can be argued, of course, that moulded notches are the more realistic in relation to the service performance of end-products but there seem to be only tenuous links between laboratory data and practical experience, largely because of the great variety of service items, production routes etc. Some aspects of moulded notches are discussed in Supplement S11.8.

11.9

In a Charpy flexed beam test, impact energies of about $1 \, kJ/m^2$ are typical for thermoplastics in a brittle state and energies of $10–40 \, kJ/m^2$ are typical for them in a ductile state. Following Vincent[7], it was once widely thought that impact resistance could only be quantified approximately. He suggested classification into four groups, for impact performance over a range of temperatures, viz.

1. Brittle when unnotched
2. Charpy impact energy $<20 \, kJ/m^2$ for notched specimen with 2 mm notch tip radius
3. Charpy impact energy $<10 \, kJ/m^2$ for notched specimen with 0.25 mm notch tip radius
4. Charpy impact energy $>10 \, kJ/m^2$ for notched specimen with 0.25 mm notch tip radius

In view of the relatively poor correlation between impact resistance as measured arbitrarily in a laboratory and as manifest by a service item, it seemed then that a relatively coarse system of classification such as Vincent's would have to suffice. Indeed, with this approach remarkable progress was made in the measurement of and understanding of impact resistance. However, the introduction of instrumented impact tests has provided enhanced quantitative insight into what happens during a laboratory impact event. This has also been enhanced by more recent developments such as low incident energy impacts and simultaneous photography of the test piece during excess energy impact. It is now possible to apportion fractions of the absorbed energy to the various stages of the failure and to separate physically meaningful features in the observed response from inadvertent ones, but see Section 11.11 and Supplement S11.11.

11.10

In the use of instrumented impact techniques for the routine assessment of impact resistance, analysis centres upon the peak force or the peak stress, the energy or the energy per unit cross-section to the peak and the energy or the energy per unit cross-section to complete failure. This is despite growing evidence that those easily identifiable quantities do not necessarily relate directly and uniquely to important stages in the development of impact damage. For research purposes and for assessment in relation to

Table 11.10/1 Comparative data on impact resistance measured by the flexed plate method, temperature approximately 20 °C

Material[a]	Impact velocity (m/s)[c]	Test piece thickness (mm)	Maximum force (N)	Energy to peak (J)	Total failure energy (J)
Carbon fibre/PEEK	4.43	2.12	6600 (510)[b]	10.3 (1.3)	42.4 (2.5)
Carbon fibre/epoxy	4.43	2.11	5130 (85)	6.6 (0.8)	30.5 (1.4)
Poly(ethersulfone injection mouldings	4.43	2.09	6870 (150)	75.4 (7.1)	75.4 (7.1)
HDPE ('Rigidex' 075-60N) injection mouldings	3.0	2.02	2440 (156)	15.9 (3.1)	
Propylene-ethylene copolymer (8% ethylene) injection mouldings	3.0	1.57	2052 (76)	10.9 (1.7)	11.4 (1.4)
Polycarbonate injection mouldings	4.43	1.92	6890 (255)	75.9 (5.3)	78.1 (6.0)

[a] Properties vary from grade to grade and therefore data should be qualified by adequate sample identification.
[b] Standard deviations in parenthesis.
[c] Moves towards international standardization have largely eliminated disparities in test configuration and impact velocity.

critical downstream applications, the force associated with initial damage and the energy absorbed up to that point may be preferred.

Despite such ambiguities, instrumented impact data nevertheless provide more quantitative information about impact resistances than is available from uninstrumented tests. Not least, they provide a view of the shape of the force-displacement signal. Table 11.10/1 is an example of flexed plate data that are now widely available and by which plastics materials can be compared, though that is not a straightforward operation when the materials being compared fail in radically different ways as, for instance, the carbon fibre composite and the polycarbonate do. The former breaks by a splintering delamination whereas the latter fails by ductile penetration, see Fig. 11.10/1. Useful inferences can be drawn from the relative values of the three characterizing quantities that are tabulated. For example, the ratio between the total failure energy and the energy to the peak, or between their normalized energies per unit cross-section, in the response curve gives an indication of the level of ductility, the propensity for the material to work harden and anisotropy of the test piece, provided that due allowance can

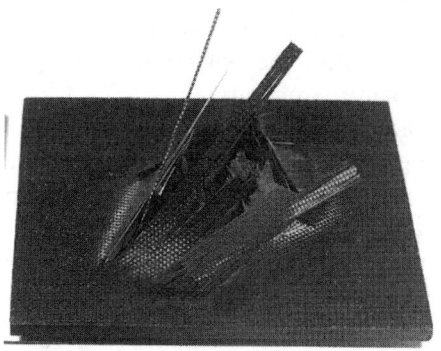

11.10/1. Impact failures in a carbon fibre composite and a
polycarbonate. In isolation, the energies entailed in total failure and
other quantities may give only a partial picture of the relative impact
resistance of materials.

be made for extraneous energy-absorbing mechanisms after the test piece
has yielded. The ratio approaches unity as the test temperature is lowered,
thereby offering an arbitrary definition of the tough/brittle transition temperature similar to that available from analogous ramp-excitation tests.

Similarly, the ratio of peak energy to peak force should be almost constant in the brittle region and a tendency for it to rise with increasing test temperature possibly signifies the appearance of some ductiliy, though viscoelasticity may be contributing also.

In one respect impact data from flexed plate tests are less convenient to use than those from flexed beam and tensile tests. The purely geometric differences in the test pieces (i.e. neglecting through-thickness differences in molecular state, etc.) cannot be eliminated by normalization via the cross-sectional area because the relevant stress equations are inconveniently complicated. Peak force, energy to peak, etc. remain as functions of the thickness and valid comparisons cannot be made between results from test pieces with radically different thicknesses. It transpires, however, that there is an interesting pattern in the effect of thickness on which pragmatic adjustments of the data can be based, see Supplement S11.10.

11.11

Evidence about the initiation of damage during an impact and the early course of damage development is usually destroyed or obscured by subsequent events in excess energy tests. Intensive studies by low energy impacts, mainly in the flexed plate configuration, and by flash photography of the test piece at a preselected moment during an impact event have led to the following conclusions:

1. The onset of fracture is not necessarily associated with a readily identifiable feature of the impact response curve.
2. In many instances, but not all, the maximum force is not directly related to a particularly significant stage of impact failure.
3. The total energy absorbed during an impact may include a large component that is unconnected with failure *per se*.
4. Electrical filtering of the response signal may apparently simplify the interpretation of the observed response by suppressing extraneous vibrations but it may simultaneously obscure meaningful features.
5. Fibre-reinforced materials are particularly amenable to investigation by low energy blows because at low incident energies the fibres tend to inhibit catastrophic crack propagation.

See Supplement S11.11 for some of the evidence.

11.12

The variety of failure classes that are observed in one material over a temperature range or sometimes at a single temperature if that lies in the ductile-brittle transition region pose interpretation difficulties to a data utilizer. Bimodal frequency distributions of failure energy with one peak associated with brittle failures and one with ductile failures are common, and multimodal distributions also arise (Section 11.6). Even with a relatively simple bimodal distribution, mean value and standard deviation verge on the meaningless and reliable comparisons between materials then entail the testing of many test pieces, which is a reversion to the use of multiple tests that was common before the advent of instrumented tests. Similarly, the presence of an outlier in a set of data raises questions as to how the data should be interpreted and what further experimentation may be required.

When the failures in a subset of specimens are all similarly brittle or ductile, the matter is superficially straightforward. A good estimate of the mean failure energy can be derived from, say, ten specimens but even so such a sample gives no reliable indication of the chance of an occasional departure from the norm. Outliers can be rejected from the quoted results if they satisfy a certain criterion of deviation or they can be incorporated into the set by use of median rather than mean as a statistic. Even so, it is desirable that any low-value outlier should be investigated since it may signify a likelihood of occasional, but possibly important, failure in service. A first step in such cases, before statistical processing and/or diagnostic investigation, could profitably be the rapid testing of a large number of specimens at some low incident energy, with the only consideration being the incidence of brittle failure; uninstrumented flexed plate tests are ideal for such a purpose, largely because such tests can be carried out very expeditiously.

11.13

The relative ease with which any impact test can be conducted belies the complexity of the phenomena and the tenuous nature of some of the inferential paths relating laboratory test data to meaningful assessments of impact resistance and serviceability. Apart from the mismatches between notional and realized excitations, between impacts in service and those in an idealized, formal test, and between the behaviour of individual test pieces and that of the entire population of test pieces, test results can be very sensitive to even minor changes in the composition of a material, to the state of molecular order in the test piece after the processing stage, to the shape of the impacted item and to other features. Such potential sources of variation are troublesome for an experimenter in some circumstances but are exploitable, with impact tests in variety being widely used to monitor progress in product development programmes, etc., as is evident from the numerous published papers that each year record the use of impact tests as a principal indicator of the benefits (or otherwise) of new additives, compounds, polymerization routes, etc. Impact test results serve also as an indicator of general product quality via lower than expected values of the characterizing quantities, untypical brittleness and raised tough/brittle transition temperature. Those deficiencies can often be detected also by other mechanical tests but impact tests tend to be more effective for the following reasons:

1. They entail high straining rates and thereby induce a lower ductility in the test pieces at any particular temperature; they can often thereby indicate the proximity of a ductile–brittle transition when that is not obvious from other classes of data.
2. They can be carried out expeditiously and, even without quantitative analysis, the results are usefully indicative of the ductility or brittleness of the test piece, especially when, as is increasingly the case, there is an archive of comparable data that can constitute a frame of reference.

Thus, impact data are widely used to support judgements about quality and fitness-for-purpose. In many such cases, the test method can be an arbitrary one provided the results are known to correlate with satisfactory serviceability but Standard Methods or other widely accepted procedures should be used wherever possible to avoid a proliferation of data classes and to give quantitative credence to the data and to the judgements based thereon.

11.14

The final and almost accidental consequence of the ubiquitous use of impact tests has been an overwhelming proliferation of data but *en masse* they have been far from uniform. Test procedures have not always conformed to a Standard Method. The method and the sample state have not always been recorded. Test programmes have often necessarily been designed and conducted with a specific and narrow objective in mind. Corrective measures in the form of standard recommendations on the presentation of data are too recent in inception to be effective yet.

The main impediment to rapid progress towards uniformity in data formats is a fundamental difference between the *modus operandi* for the generation of data in relation to plastics materials and that for the generation of data in relation to plastics end-products. There are several important differences in the data requirements of the two groups (and also regions of overlap). With some simplification, it can be said that the first group would seek to measure pertinent physical properties over a wide range of the influential variables whereas the second group would prefer to measure attributes over a selectively narrow range of the variables relevant to a specific service situation. Thus, most of the contents of a database emanating from the first group might be deemed irrelevant by the second group and what is pertinent might be deemed to be insufficiently detailed. The situation is similar in some respects to that relating to creep databases, see Supplement S6.14. There are some corresponding comments on impact resistance databases in Supplement S11.14.

One important bridging step was the recent development of a standard for the measurement of impact toughness. The European Structural Integrity Society (ESIS) in its task committee on polymers and composites has successfully developed a measurement method for fracture toughness (K_C and G_C) at an impact velocity of 1 m/s. A single edge notched beam is used and the analysis is identical to that previously discussed for linear elastic fracture mechanics. The particular problem attendant on conducting this measurement at 1 m/s is that fracture will often be complete within 1 millisecond, particularly for brittle materials. That usually causes extraneous vibrations to be superposed on the force–time signal, sufficient to seriously limit the detection of peak force and peak energy. The ESIS approach enables mechanical damping to be applied to the specimen resulting in a change in the contact stiffness between the impactor and specimen sufficient to allow adequate interpretation of the signal, but not entailing so much mechanical damping as to obscure important features. Corrections for the identation of the impactor on the specimen and the compliance of the test machine need to be made in order to accurately determine both K_C and G_C.[9]

References

1 J. J. Mooij. 'Instrumented flat-headed falling-dart test.' *Polymer Testing*. 2, 69, 1981.

2 P. E. Reed, P. J. Nurse and E. H. Andrews. 'High strain rate testing of plastics. Part 1. Shock tube technique.' *Journal of Materials Science*, 9, 1977, 1974.

3 K. Thomas and D. E. Meyer. 'Study of glass-fibre-reinforced thermoplastic mouldings.' *Plastics and Rubber: Processing*, 1, 99, 1976.

4 T. G. Meijering. 'The evaluation of the ductility of thermoplastic pipes with an instrumented falling weight test.' *Plastics and Rubber: Processing and Applications* 5, 2, 165, 1985.

5 B. Terselius, J.-F. Jansson and J. Bystedt. 'Gelation of PVC, Part 4: impact strength.' *Plastics and Rubber: Processing and Applications*, 5, 1, 1, 1985.

6 E. Plati and J. G. Williams. 'The determination of the fracture parameters for polymers in impact.' *Polymer Engineering and Science*, 15, 6, 470, 1975.

7 P. I. Vincent. *Impact Tests and Service Performance of Thermoplastics.* A Plastics Institute Monograph. UDC reference number 678.073.531.66. First published 1971.

8 G. Merle, O. Yong-Sok, C. Pillot and H. Sautereau. 'Instrumented and temperature-controlled Charpy impact tester.' *Polymer Testing*, 5, 37, 1985.

9 A. Pavan and S. Draghi. 'Further experimental analysis of the dynamic effects occurring in three-point bending fracture testing at moderately high loading rates and their simulation using an *ad hoc* mass spring model.' Proceedings of 2nd ESIS TC4 Conference on Fracture of Polymers, Composites and Adhesives, ESIS Pub 27, p347, Elsevier, 2000.

S11.4 Practical aspects of conventional impact test equipment

General

The three common types of impact test are simple in concept and the early apparatus was relatively crude, reflecting a contemporary simplistic attitude of 'just bash it and see if it breaks' on the part of the experimenters. However, the emergence of instrumented tests and the associated potential for enhanced data initiated what became a continual process of refinement in the cause of greater precision.

This Supplement is a summary of those practical aspects of the tensile and flexed beam and flexed plate impact test machines that can affect the precision, accuracy and validity of the data generated. This may influence an investigator's choice of apparatus for the resolution of a particular issue.

The engineering design of the machines and the fine details of the devices are to be found in manufacturer's literature and practical test manuals. Here, the functioning of the machines is considered as the sequence – delivery of the impulse, the stress state established in the impacted item, the response of that item and the utility of the information contained in that response. In those respects, the test methods differ significantly and are differently suitable for particular tasks, which explains the continuing widespread use of the three configurations.

All impact test machines have one characteristic in common – they measure certain quantities related to the impact resistance of the impacted item, which is not necessarily closely related to the impact resistance of the material from which the item is made. The trains of inference leading back to the impact resistance of the material (if such a concept is meaningful) vary with the circumstances and differ from test configuration to configuration.

Delivery of the impulse

The standard versions of the tensile and flexed beam machines employ a pendulum to deliver an impact. In contrast, the flexed plate machines usually employ a falling mass. Irrespective of the method, structural considerations tend to limit the energy that can be delivered via pendulum or falling mass, and consequently in some instances the impactor decelerates markedly during the impact event and the excitation function degenerates from the ideal impulse. Also, the impact velocity of the falling mass devices may be affected by friction in the guidance device that has to be used to ensure a specific point of impact; corresponding energy losses occur at the fulcrum of a pendulum.

Large, tough test pieces require large, robust machines and hydraulically driven darts (impactors) can usually be designed to deliver adequate energies, but they tend to accelerate slowly and not necessarily reproducibly to the desired velocity. Those various deviations from the intended impact velocity can, of course, be monitored, but only at the cost of additional complications in the equipment.

Tensile impact entails an additional mechanical link between the energy source and the specimen, to translate the movement of the actuator (pendulum, falling mass or driven dart) into a tensile deformation. That can be a source of additional noise on the response curve, compared with impact by the other methods, but the delivery of the impulse is not otherwise affected.

The stress state established in the impacted item

In principle, a tensile stress state should be the easiest of all to attain because local distortion of the stress field at loading points can be eliminated by the use of appropriately profiled specimen end-sections. However, unless the grips are symmetrically assembled on the test piece some bending is likely to be superposed onto the tensile deformation. Additionally, if the actuator is a swinging pendulum, and especially if the machine is a typical small bench model, the moving grip will follow a curved path and the specimen will bend accordingly unless the two grips are completely free to rotate, which is an idealization that is virtually impossible to attain.

Flexed beams are next in simplicity. Distortions of the stress field should only occur at loading points. However, the clamped cantilever geometry of the Izod test is more troublesome than the simple supports of the Charpy test because a practical cantilever falls short of the mathematical idealization and results also depend on the clamping pressure. If beam specimens are small, as in the standard versions of the tests, the stress field distortion may extend over a significant fraction of the span, rendering the simple (elastic) beam equations correspondingly inaccurate. Another source of error, which arises also in the flexed plate configuration, is surface tractions at the points of contact between the test piece on the one hand and the supports and the impactor on the other. The surface tractions are unimportant if the failure is brittle but they can affect the peak force and the total failure energy if the failure is ductile.

The stress field in flexed plates is more complicated overall. First, test results cannot be readily converted into stresses so that whilst flexed beam and tensile test results are easily, if sometimes inaccurately, expressible as stress and in energy per unit cross section, results from flexed plate tests have to be expressed as force and energy and are only directly comparable one with another if the test pieces are of the same thickness, see also Supplement S11.10. Secondly, as a crack or tear develops, its tip moves into a less severe stress field and may cease to grow, even though the impactor passes through the specimen. This generates an optimistic but erroneous impression of resistance to crack growth. In contrast, if the crack or tear remains in the region of the stress field most favourable to crack opening, a pessimistic impression of resistance to crack growth may be inferred.

Notches change the stress field in the test piece and sometimes radically reduce the impact resistance by increasing the stress triaxiality. They are readily accommodated in the tensile and the flexed beam configurations, subject only to the limitations inherent in the theory of fracture mechanics. The flexed plate configuration suffers the complication that a linear notch in a biaxial stress field is less amenable to analysis; however, axisymmetric stress concentrators, primarily holes or part-through holes at the point of impact[1], seem to be sufficiently efficient as embrittling features to discriminate between tough polymers (for instance, under the identical conditions of flexed plate and axisymmetric notch polyethersulphone is embrittled whereas polycarbonate remains ductile).

Apart from such specific advantages and disadvantages, the choice of test method is often governed by the severely practical issue of the convenience of specimen preparation. For example, it is possible to see advantage with the flexed plate method because the standard test requires either discs 60 mm in diameter or squares of side 60 mm, the edge condition of which is unimportant, whereas the flexed beams have to be rectangular parallelopipeds of a specified size and the tensile bars have to be machined or directly moulded to a standard profile and with high-quality surfaces. Additionally, at one time, oversize flat discs were commonly impacted in the flexed plate configuration to provide insights into the influence of surface features and surface quality on the impact resistance without the cost and delay of a machining stage.

That non-standard practice is still resorted to and the effect of the overhang has been studied. In the case of an amorphous plastic in its glassy state (PMMA with diameter close to that of the support ring) the response curve rises to a peak after which the force falls quickly to zero as the crack propagates to the edge of the support beyond. As the size of the test piece is increased relative to the support ring, the first part of the curve up to the first peak is unchanged, but the force does not fall immediately to zero because once the crack reaches the edge of the support it slows down since the stress field outside the support is less severe. It thus takes longer for the crack to reach the edge of the test piece, longer for the impactor to penetrate it, and longer for the force to fall to zero. The extra energy absorbed is a function of the test piece geometry rather than a reflection of the material properties. The peak should be taken as the criterion of failure, though not of fracture initiation which occurs much earlier in the impact event. In the case of a tough material above its glass–rubber transition (polyethylene and polypropylene at room temperature) with a test piece size close to the diameter of the support ring the impactor simply pushes through. As the size is increased, the energy required for penetration decreases as does the deflection.

The overall conclusion is that, if oversized specimens are to be used regularly a standard overhang should be adopted, preferably between one-third and one-half larger than the support ring, to avoid misleading comparisons between samples with different overhangs, and the use of such oversize test pieces should be limited to local contexts.

Directly moulded discs of 60 mm diameter are allowed by the standard specification and can obviously provide the same facility for rapid comparisons as the use of oversize test pieces, but the data have to be viewed with some reservations because of likely interactions between that small cavity size and the gate geometry causing atypical flow geometries and irregularities.

Notches modify the stress field both directly through their geometry and indirectly by surface imperfections on machined notches and flow irregularities near moulded notches. The quality requirement for machined notches is similar for all test configurations. For moulded notches it is essential that there should be no flash because the molecular alignment in the flash or in the neighbouring body of the moulding can be so extreme that it constitutes a starter crack that may give grossly misleading indications of the notch sensitivity. On balance, data from flexed plate tests are less likely to be sensitive to flash than data from the other configurations. It has to be noted here that data from specimens with moulded notches should always be viewed circumspectly because the response probably reflects effects of the flow geometry in addition to those of the stress concentration. However, when the potentially detrimental effects of features such as grooves, bosses, holes, etc. have to be evaluated, the flexed plate configuration is usually the more versatile and otherwise superior method, even if the test pieces are oversize.

The response to the impulse and utilization of the results

Instrumentation of the impact equipment provides, usually, a force–time signal of the impact event. That record is the response of the entire structure which comprises test piece and test machine. The response of the test piece has then to be either inferred or extracted by judicious filtering. Filtering is a risky process that may delete more than unwanted signals. A ductile test piece acts as a damper on the elastic vibrations of the system and the recorded response curve may then be relatively featureless. A brittle test piece exerts no such damping influence and the observed response is a combination of test piece and machine vibrations. Another factor that affects the degree of damping is the condition of the surface of the specimen arising from the processing stage, from subsequent service and from inadvertent contamination, e.g. moisture.

The tensile and flexed beam methods enable the investigator to assess impact resistance along a defined axis, whereas the flexed plate method largely overrides such choice and automatically identifies the weakest direction in the plane. Assessments of the anisotropy may therefore require the directional specificity of the tensile and flexed beam methods whereas the flexed plate method is likely to be the more efficient in identifying the lower bound of performance.

There has always been a widely-held view that impact testing is inherently imprecise and that the impact resistance of a material or an end-product is highly variable, which has fostered the belief that the quantitative value of impact testing is severely limited. At least two factors may have contributed to that view. First, impact testing was a relatively rough and ready, rapid procedure that was inevitably prone to corresponding possibilities for error. Second, impact tests were often used, and still are, to explore the boundary between acceptable ductile behaviour and unacceptable brittle behaviour, which entails tests in the ductile–brittle transition region where the coefficients of variation (c.v) are often large (Section 11.7). The present reality is that properly conducted instrumented impact tests on good-quality samples and test pieces can yield results with similar coefficients of variation to those from most other mechanical tests; thus, with best practice, the coefficient of variation quantifying the precision is approximately 0.10 for brittle failure, possibly as little as 0.03 for the onset of ductile failure but certainly much more for mixed-mode failure in the ductile–brittle transition region. The coefficients quantifying accuracy are probably similar provided that extraneous factors such as interspecimen and intersample variations are eliminated or excluded from the computations. However, the reliability of those estimates, the relative precision and accuracy of the data generated by the various test configurations and the reliability of many of the data quoted in the literature cannot usually be reduced to simple statements because the qualifying data and the experimental details are absent from many of the published papers in which impact tests have merely been a means to an end such as the development of an improved grade of a material.

In relation to that last comment, it is salutary to consider some data in one of the more informative papers[2] that was outstanding in its wealth of experimental results. Table 6 of that paper contains notched Charpy data relating to 37 different PVC compounds variously containing two types of soft filler (at 10 parts per hundred of the resin), 15 different compatibilisers and other minor differences in the constituents; four of the compounds were controls containing no soft filler. The test pieces were machined from compression moulded sheets, carried out in accordance with ISO R179. The mean values of 'impact strength' in each set ranged from 1.4 to 17.1 which were associated with coefficients of variation of 0.143 and 0.064 respectively, which was in the order that one might expect for brittle and ductile failures respectively. The c.vs of the 37 sets ranged from 0.007 to 0.294 and the mean value of the distribution was 0.154. One possible explanation as to why these variabilities are much higher than those suggested as the best attainable is inadvertent variations in the compounding and further processing stages. Similarly, coefficients of variation that have been published elsewhere tend to be larger than the ideal expectations, but they differ from paper to paper, presumably due to the distorting influence of operational factors and peculiarities of each class of material. There is no clear evidence of machine-specific influences provided that known factors, such as clamping pressure in the Izod configuration, are under control.

On balance, and as one might reasonably expect, it seems that the three configurations yield similar levels of precision. The matter of relative accuracies cannot be resolved even approximately because the disparities between the measurable quantities, peak force, energy to break, etc., are complicated functions of the material and the test configuration to the extent that each test method measures particular self-defined 'properties'. That vagueness carries certain implications for data quality, for the process of product design and for the scope and format of impact resistance databases.

References

1 A. Martinez, P. E. Reed and S. Turner. 'Flexed plate impact testing of poly(ether sulphone). Part 2 Notched specimens.' *Journal of Materials Science*, 21, 3801, 1986.
2 F. H. Axtell, K. Jarukumjorn and W. Sophanowong. 'The influence of interfacial agents on the impact strength of poly(vinyl chloride) and soft filler composites.' *Plastics, Rubber and Composites: Processing and Applications*, 22, 2, 79, 1994.

S11.5 Analysis of impact response

Ideally, an instrumented impact test should provide a record of force and simultaneous displacement throughout the course of the event, but accurate measurement of the displacement is virtually impossible at high rates of change. A common solution is measurement of force as a function of time and calculation of velocity, displacement and energy. This is achieved by describing the force (P) on the striker of mass m at time t, as shown in Eq. S11.5.1:

$$-P = m\frac{dv}{dt} - mg \qquad [S11.5.1]$$

By integration an expression can be obtained for the velocity of the impactor (v_t):

$$v_t = v_o + gt - \frac{1}{m}\int_0^t Pdt \qquad [S11.5.2]$$

where v_o is the initial velocity of the striker.

Further integration of Eq. 11.5.2 provides the displacement of the test piece at any time (x_t):

$$x_t = v_0 t + \frac{gt^2}{2} - \frac{1}{m}\int_0^t \left(\int_0^t Pdt\right)dt \qquad [S11.5.3]$$

The energy (U) can then be written as:

$$dU = Pdx = P\frac{dx}{dt}dt = Pvdt \qquad [S11.5.4]$$

Therefore the energy absorbed by the test piece at any time (U_t) can be obtained by combining Eqs. S11.5.2 and S11.5.4:

$$U_t = v_0\int_0^t Pdt + g\int_0^t Pdt - \frac{1}{2m}\left(\int_0^t Pdt\right)^2 \qquad [S11.5.5]$$

By this method the experimental errors that would arise in the measurement of displacement are avoided but they are replaced by accuracy limitations inherent in the repeated integration process.

The usual arrangement is for the force transducer to be sited in the impactor. Although the force–time data directly relate to the impactor it is assumed that they also relate without change to the test piece. Therefore, the striker and specimen are assumed to be in contact throughout the impact event and the specimen is assumed to react elastically, which is seldom strictly true even in the early stages. Additionally, the local stress field in the immediate vicinity of the impactor is far from simple, particularly for the flexed plate configuration. Similarly, the strain is not known accurately because of errors arising either from the integration process or from the stress field calculations. It follows that a calculation of the energy density at the point of impact, which is the critical quantity, is likely to be completely unreliable. Computed values of total energy should be more reliable, even though they must reflect the uncertainties associated with possible extraneous contributions to the total energy.

In practice, the force is usually displayed as a function of time rather than as a function of the computed displacement. This is adequate for most diagnostic purposes, especially when the incident energy is far in excess of the failure energy and the displacement is consequently almost a linear function of time. That simplification is invalid for low incident energy tests but even for those the force–time record may suffice if only because the calculations are likely to be only approximations under those conditions because of other influential factors. The disparity between the two displays may be represented by the approximate relationship equation S11.5.6, which is widely used and which was assessed by Bevan et al.[1]

$$U_{Px} \approx U_{Pt}\left[1 - \frac{U_{Pt}}{4U_A}\right] \qquad\qquad [\text{S11.5.6}]$$

where U_{Px} = energy calculated as the area under force vs. displacement
U_{Pt} = energy calculated as the area under force vs. time
U_A = the incident energy.

Reference

1 L. Bevan, H. Nugent and R. Potter. 'The excess energy approach to the impact testing of plastics.' Polymer Testing, 5, 3, 1985.

S11.7 Prediction of coating thickness to avoid brittle fracture

Perhaps the most common scenario to address for fracture of a coating material relates to the possibility that it can behave in either a ductile or a brittle manner, dependent on the service conditions. For substrates that are coated, it will be important to design the laminate in such a way that only ductile fractures in the coating material can be accommodated.

Impact loading of a laminate on the non-coating surface is such a case. The requirement is for the coating to be applied at such a thickness that brittle cracks do not occur over a range of laminate temperatures. There are four considerations:

1   Will the coating fracture in a ductile or brittle manner?
2   How will the thickness of the coating influence this?
3   How will temperature influence this fracture scenario?
4   Can one predict when a transition will occur from ductile to brittle fracture in the coating material?

As an example these issues are considered for a polycarbonate substrate with an epoxy coating.

The general case for ductile–brittle transitions in the coating material is given in Fig. S11.7/1. An absorbed impact energy vs. temperature view is shown in the top diagram, where it can be seen that low energy absorption accompanied by brittle fracture will occur at low temperatures. On the other hand, high energy absorption with ductile fractures will occur at high temperatures and a ductile–brittle transition will occur at some intermediate temperature. We now require to translate this to an energy versus coating thickness profile at some fixed temperature and therefore construct the lower plot in Fig. S11.7/1.

Brittle fractures can be expected in the coating material when the coating thickness is large. This is because the thickness will be large compared with the plastic zone around any inherent flaw in the coating material. This maps the low energy and high thickness portion of the sketch. As the thickness of the coating becomes decreasingly smaller, a stage will be reached when the plastic zone size around an inherent flaw will be comparable with the coating thickness. For such conditions, ductile fracture can be expected. This maps the high energy and low coating thickness portion of the sketch. Therefore, a ductile–brittle transition in the fracture of the coating material can be expected between these two regions at an intermediate thickness. The lower diagram in Fig. S11.7/1 shows the full shape of an energy vs. coating thickness plot at a specific temperature.

In the energy vs. thickness diagram, it is known that the thickness of the coating at the start of brittle fractures will be at least an order of magni-

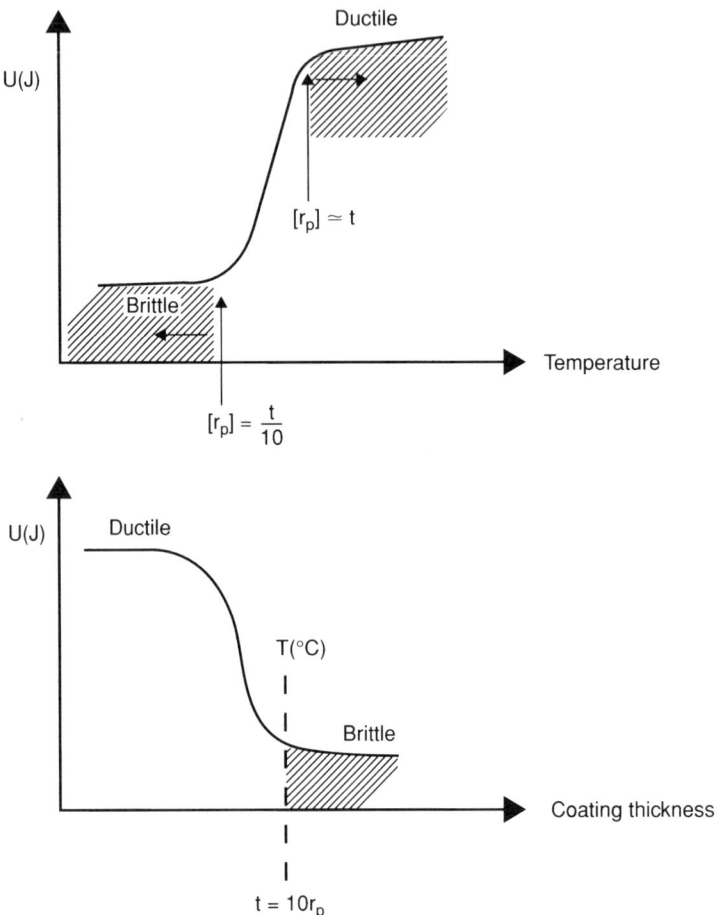

S11.7/1. Criteria for ductile–brittle transitions in coating material. Top curve on the basis of energy vs. temperature. Bottom curve on the basis of energy vs. coating thickness.

tude larger than a plastic zone size. Moreover, for impact conditions of the laminate it can be expected that unconstrained deformations will occur. Therefore, we can associate the fracture with plane stress conditions for the plastic zone radius (r_P):

$$r_P = \frac{1}{2\pi}\left(\frac{K_C}{\sigma_y}\right)^2 \qquad \text{[S11.7.1]}$$

where K_C is a plane strain value for fracture toughness and σ_y is a tensile yield strength.

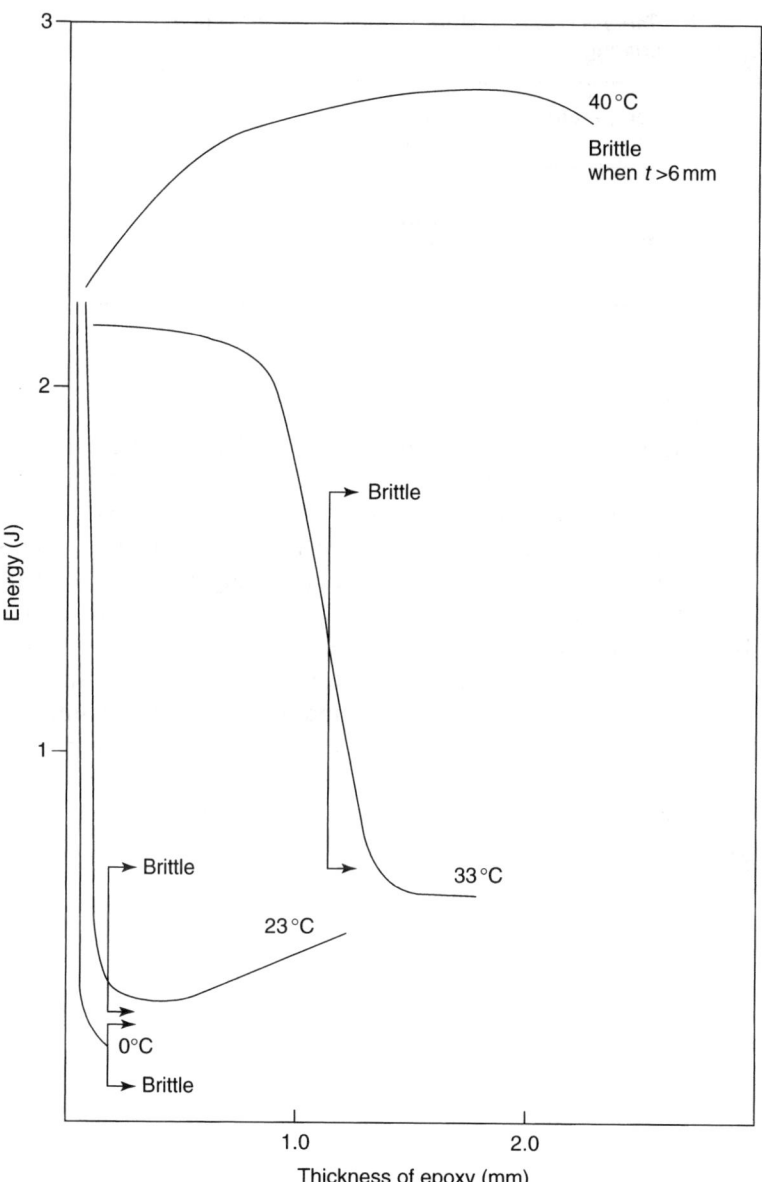

S11.7/2. Energy vs. coating thickness at various temperatures where a
comparison is made between predicted and experimental
ductile–brittle transitions.

Table S11.7/1 Fracture properties for the epoxy-coating material

Temperature (°C)	K_C (MPam$^{1/2}$)	Yield strength (MPa)
0	0.9	82
23	0.82	71
33	1.64	61
40	3.11	51

With a knowledge of those two material properties for the coating material, it would be possible to calculate the plane stress plastic zone radius and from that determine the thickness condition for a ductile–brittle transition in the coating material, when the laminate is impacted on the substrate side. If that is known at one temperature, it can also be made available at a range of temperatures and hence the temperature dependence of the laminate in impact can be predicted. Table S11.7/1 summarizes the fracture properties for an epoxy-coating material for a range of temperatures.

Impact data on the laminate structure for a range of coating thicknesses has been obtained and can now be superimposed with the prediction of the coating thickness where a ductile–brittle transition will occur. These results are shown in Fig. S11.7/2 and the agreement between measured and predicted coating thickness for the ductile–brittle transition is encouragingly good.

S11.8 Moulded notches and impact resistance

Many features of the flow path via which a molten thermoplastic is transformed into a solid object affect the properties of that object. The degree
to which each feature is influential varies with the rheological characteristics of the melt as affected by its temperature, pressure, shear history etc.
Similarly, in fibre composites there may be regions starved of reinforcing
fibres or containing a disproportionate volume fraction of fibres that are
aligned unfavourably. Fortunately, considerable knowledge of these matters
does now enable loss of toughness in an end-product to be minimized,
though the complexities in practical manufacture and processing cannot be
fully eliminated. In many cases, detrimental flow geometries, flow irregularities and structural inhomogeneities act as stress concentrators and
therefore estimates of 'notch sensitivity' as a simplifying concept are frequently required because the various polymer classes are differently sensitive to stress triaxiality and the local anisotropies.

Thus the impact testing of notched test pieces was commonplace long
before instrumented tests were developed and has continued with the addition of fracture mechanics analysis for sharply notched pieces. In the particular case of injection-moulded notched Charpy bars, which are often used
in the interests of economies in specimen preparation time, differences in
flow length and gating can cause significant variations in the outcome of
tests. In one experiment,[1] for instance, there were five large sets of specimens (30 in each set) moulded in a two-cavity and a three-cavity mould.
The two cavities in the former were nominally identical end-gated bars
for which the mean values of the energy to fail (and the standard deviation
in parenthesis) were $6.2(0.6) \, kJ/m^2$ and $6.0(0.4)$, which are virtually identical. The three-cavity mould also had an end-gated bar but the mean impact
energy was significantly lower at $3.4(0.5)$, probably because the gate area
was double that for the two-cavity mould and the flow path was shorter
so that there was probably lower molecular orientation along the bar in
the vicinity of the notch. The other two cavities in that mould were side-
gated directly opposite the notch and nominally identical; the mean values
were $2.7(0.4) \, kJ/m^2$ and $2.1(0.6) \, kJ/m^2$, significantly lower than the values
for the end-gated bars, as one would expect, and sufficiently different from
each other to suggest that there may have been turbulence near the entry
point.

An important point that arises following such results is that Charpy data
often have to be compared with supposedly corresponding data from other
sources. Although they are all likely nowadays to be classified by information about the processing conditions and storage history, it is virtually
unknown for the metadata to include information about the flow path, gate
size, etc. Therefore extraneous and unnotified differences in flow path such

as those descibed above may lead to false comparisons and misleading conclusions.

Similar complications arise with moulded notches in the flexed plate configuration. Figure S11.8/1 is a photograph of an injection-moulded disc typical of the simple shapes that were commonly impacted in the era of the staircase method. The impact resistance varies from point to point in the disc[2] and with the moulding conditions. There is a complex interaction between the flow paths and the alignment direction of grooves/notches in such plates and this can be studied with mouldings from cavities such as that shown schematically in Fig. S11.8/2. It is an edge-gated disc 114 mm in diameter and approximately 6.4 mm thick with an insert such that the final moulding has a diametral v-shaped groove with a tip radius of approximately 1 mm. The groove can be rotated to lie at any angle with respect to the diametral line passing through the gate and in what follows '0° groove' designates a groove lying along that particular line.

Table S11.8/1 shows some interesting results for such discs moulded from three grades of poly(ether sulphone).[3] In the standard test, i.e. impact of a flat smooth plate, unnotched specimens of poly(ether sulphone) are ductile;[4] if the impactor has sufficient energy, it penetrates the specimen neatly and

S11.8/1. Injection-moulded edge-gated disc with surface grooves. Such discs have been widely used as simulations of plate-like service components, but the impact resistance depends strongly on the flow length and the gate geometry.

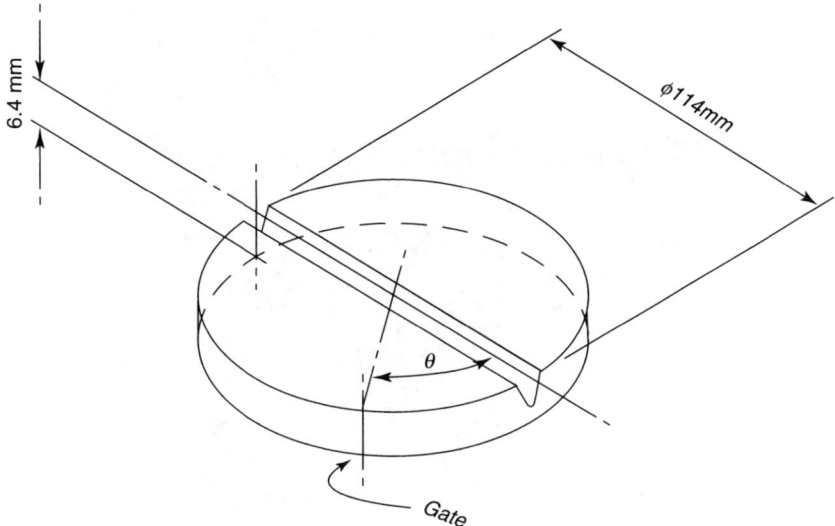

S11.8/2. Diagram of grooved disc used in studies of the interaction between notch and flow path. θ is the angle between the main flow axis and the axis of the groove.

Table S11.8/1 The flexed-plate impact resistance of discs with a moulded notch; injection-moulded poly(ether sulfone) at 23°C

Test sample	Groove alignment	Peak force (N)	Energy to peak (J)	Total energy (J)
Commercial	0°	2762 (617)	9.6 (4.5)	16.5 (4.9)
Grade A	90°	2922 (401)	4.1 (1.5)	8.0 (0.9)
Commercial	0°	4822 (1211)	13.1 (8.2)	23.1 (11.6)
Grade B	90°	4560 (999)	6.0 (3.4)	12.4 (3.1)
Experimental	0°	3113 (721)	9.5 (5.4)	17.8 (8.0)
Grade	90°	2399 (185)	2.4 (0.5)	5.5 (0.5)

a typical value for the energy to the peak is 75 J for a specimen 3.2 mm thick and 130 J for one 6.4 mm thick. In contrast, if the specimen has been sharply notched in the tensioned face directly in line with the point of impact, the fracture is brittle and the energy required is only a few joules. The moulded notch depicted in the figure also induces brittleness, as can be inferred from the magnitude of the property values in the table and from the photograph of the fracture surface in Fig. S11.8/3. That photograph also shows local plasticity (indentation) at the point of loading and at two diametrally opposed points on the support ring. If the specimen had been isotropic, the

S11.8/3. Tension surface of broken grooved disc (half) showing brittle fracture surface and severe indentation due to reaction forces at the support ring. The reaction was not uniform along the line of the support because the grooved disc was anisotropic in stiffness; impact at the centre of the specimen induced incipient hinging along the groove.

reaction at the support ring would have been uniformly distributed along a corresponding circle but the groove confers a lower stiffness in the orthogonal direction and, under central impact, the specimen tends to fold along the line of the notch, like a book but, irrespective of that detail, the energy absorbed by such plastic deformation away from the site of failure distorts the measured value of total energy and possibly that of energy to the peak.

The discs in which the main nominal flow axis lies along the notch are significantly tougher, as measured by the energies, than those in which it lies across the notch. The energy to the peak and the total energy are apparently much more sensitive to the alignment of the groove than the peak force is. The relative toughnesses are the reverse of what was found for beams cut from the discs and also of what would be expected from a simple analysis of the flow geometries. An anomaly might be expected to arise from the complex flow regime that exists at the groove, where the thickness restriction creates a high impedance flow path along what would be the main flow axis in the absence of a groove, thereby distorting the conventional flow geometry of an edge-gated disc of uniform thickness. However, the explanation may simply be that the absorbed energy data mainly reflect extraneous local plastic deformations such as those in Fig. S11.8/3, rather than the brittleness induced by the notch; the corresponding energy absorbing mechanisms in the Charpy tests are likely to have been much smaller.

There have been numerous examples of unexpected brittleness under impact of injection-moulded service items, where the explanation was to be found in an unfavourable anisotropy caused by the flow geometry. For example, the traditional practice of radiusing sharp corners to reduce the geometrical stress concentration can actually be detrimental via the flow geometry. Such design errors still occur, despite the greater knowledge base, because production, aesthetic and commercial considerations may out-weigh purely technical ones and therefore evaluation programmes should allow for such potential violations of expectation and databases should reflect the possibilities for detrimental combinations of influential factors.

References

1 F. N. Cogswell, E. A. Cole and S. Turner. 'Rheology and mechanical properties: a study of the influence of processing history on the impact resistance of thermo-plastics.' *Journal of Elastomers and Plastics*, 11, 171, 1979.

2 P. E. Reed and S. Turner. 'Flexed plate impact testing of poly(ether sulphone). Part III. Flow geometry effects.' *Journal of Materials Science*, 23, 1985, 1988.

3 I. N. Scrutton and S. Turner. 'Flexed plate impact testing IV. The effect of a moulded notch with poly(ethersulphone) specimens.' *Plastics and Rubber: Processing and Applications*, 6, 3, 223, 1986.

4 S. Turner and P. E. Reed. 'Flexed plate impact testing of polyether sulphone: development of a notched specimen.' *ASTM Special Technical Publication*, 936, 262, 1987.

S11.10 The effect of thickness on flexed plate impact resistance

Because of the complexities of the stress field in centrally loaded flexed plates, test results from that configuration are quoted as energies rather than as the energies per unit cross-section that are usual with the simpler flexed beam and tensile impact data. Comparisons between data sets are correspondingly hampered unless the test pieces for each set are closely similar in thickness or unless the data can be adjusted empirically to allow for differences in thickness.

It has transpired that impact energy for total failure seems to be proportional to fractional powers of the thickness with a different power law for each class of plastic. Peak force, on the other hand, is related to thickness in a simple and loosely explicable manner, even though the peak is not always associated with a significant stage in the impact event.

The facts that have emerged are:

1 Peak force is proportional to thickness to the power n, where n lies between 1 and 2, i.e.

   $$P \propto h^n.$$

2 $n = 1$ for ductile glassy amorphous plastics, i.e. those whose mechanical behaviour can be approximated by an elastic-plastic model.[1]
3 $n \approx 2$ for brittle glassy amorphous plastics, i.e. those whose behaviour can be approximated by an elastic model.
4 $2 > n > 1$ for crystalline plastics above their glass–rubber transition temperature.
5 The value of n is affected by molecular orientation in the specimen.
6 The apparent value of n is distorted by the complexity of the stress field at and near the point of impact; the effective radius of the area of contact differs from the actual radius.

Peak force divided by thickness for polycarbonate discs is given in Table S11.10/1.[2] Edge-gated, injection-moulded discs were moulded at two thicknesses, 3.3 mm and 1.6 mm, and some were thinned by machining over one face to provide test pieces at intermediate thicknesses. Peak force seems to be linearly proportional to the thickness within the limits of experimental error, as one would expect if the material around the periphery of the impactor yields. The slightly deviant behaviour of two sub-sets of specimens that had been thinned substantially is probably attributable to the removal of the particular molecular orientation typical of surface or near-surface layers in injection mouldings, which would leave the thinned test piece with 'unbalanced' layers of anisotropy. The failure was typically a simple ductile penetration.

Table S11.10/1 The effect of thickness on the impact
resistance of injection-moulded polycarbonate
at 23 °C

Thickness (mm)	Mean peak force[a] (N)	Mean peak force/ thickness (N/mm)
3.31	13 080 (305)	3952
3.3 → 2.50	9 975 (380)	3990
3.3 → 1.92	6 890 (255)	3589
1.64	6 720 (215)	4098
1.6 → 1.09	3 995 (295)	3666

[a] Standard deviation in parenthesis.

Similar results, including a drop in the P/h ratio for specimens that had
been thinned substantially, were obtained in tests on discs moulded from
poly(ethersulphone). The mean value of the ratios for eight subsets was
3388 N/mm, with a standard deviation of 170 N/mm.[1] The corresponding
values for the five subsets in Table S11.10/1 are 3859 and 220.

For polystyrene specimens cut from injection-mouldings at three thicknesses and thinned mechanically as necessary, the ratio P/h^2 was approximately constant, see Table S11.10/2 and Fig. S11.10/1, as one would expect
of an elastic material. The deviant datum points relate to the thinner mouldings in which the molecules were probably preferentially aligned parallel
to the main flow axis and the specimens were probably correspondingly
weak along the transverse direction. The results from a flexed plate test
tend to reflect the properties in the weakest direction.

Table S11.10/2 The effect of thickness on the impact
resistance of injection-moulded polystyrene at 23 °C

Thickness (mm)	Mean peak force[a] (N)	Mean peak force/ (mean thickness)^2 (N/mm^2)
6.16	1413 (47)	37
6.16 → 5.01	903 (73)	36
6.16 → 3.75	484 (34)	34.5
2.57	209 (8)	31.9
6.16 → 2.49	219 (16)	35
6.16 → 1.61	93 (12)	36
1.57	64 (5)	26

[a] Standard deviation in parenthesis.

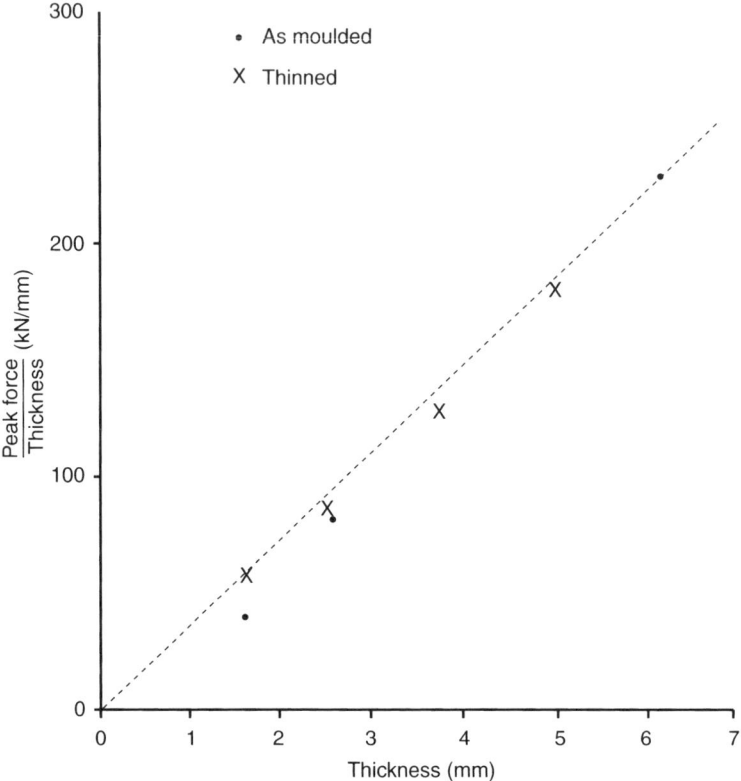

S11.10/1. The force–thickness relationship for polystyrene mouldings under standard excess energy impact. Molecules tend to align along the main flow axis in thin mouldings and the power law relationship is thereby distorted. The effective area of contact between impactor and specimen is also a function of the specimen thickness; see Eqs. S11.10.2 and S11.10.3.

Regression analysis of the data excluding the most aberrant point yields the relationship:

$$P = 33.5 \; h^{2.04}$$

where P = peak force
        h = thickness

and the correlation coefficient is 0.999.

An expectation that the power index should be exactly two for an elastic material is based on an oversimplification of the situation at the point of impact. The underlying assumption is that fracture will occur at some critical value of the stress, the expression for which is:

$$\sigma_{max} = \frac{3P}{2\pi h^2}\left[(1+v)\ln\frac{a}{r_o'}+1\right]$$ [S11.10.1]

where P = peak force
  h = thickness
  a = radius of support ring
  r_o' = effective area of contact between nose of impactor and specimen

The effective radius is taken as the true radius if the latter is greater than half the thickness, but otherwise it is:

$$r_o' = \sqrt{1.6r_o^2 + h^2} - 0.675h$$ [S11.10.2]

in which case $\sigma_{max} = \frac{3P}{2\pi h^2}[f(h)]$ [S11.10.3]

and hence peak force is not exactly proportional to thickness squared.

The rationale is helpful, even though imperfect, and provides a basis for the use of an empirical correction factor for the peak force. With certain other simplifying assumptions about the linearity of the force-displacement curve up to the peak, the energy to the peak can be similarly adjusted. Tests have to be carried out at several thicknesses.

For tough glassy polymers, the value of P/h falls dramatically if there are stress concentrators and its value reflects the severity of the concentrator.

References

1 A. Martinez, P. E. Reed and S. Turner. 'Flexed plate impact testing of poly(ether sulphone). Part 2. notched specimens.' *Journal of Materials Science* 21, 3801, 1986.
2 P. E. Reed and S. Turner. 'Flexed plate impact testing V: Injection moulded polycarbonate discs.' *Plastics and Rubber: Processing and Applications* 8, 3, 173, 1987.

S11.11   Damage development during impact

S11.11.1   Low incident energy impact

Low incident energies that generate limited damage in the impacted specimens give valuable insights into the course of development of impact damage and correlations can thereby be established between features on the response curve and the degree of damage.

Those response curves often exhibit features related to minor damage that are not clearly discernible in the response to excess energy impacts, either because of the changes that follow rapidly as the impactor continues to advance or because noise obscures the details; Figure S11.11.1/1 illustrates how low energy tests can unambiguously locate a damage event when the conventional excess energy test cannot do so with certainty. However, 'first damage' in the two cases may not be identical if the initial impact velocities are significantly different. Low incident energy is usually achieved by a lower impact velocity than that of a standard excess energy test rather than by use of a smaller mass. The velocity subsequently reduces to zero and may then reverse, with the impactor either rebounding from the specimen or remaining in contact and vibrating in unison with it.

Under appropriate incident energy conditions, limited crack growth not progressing to complete fracture can be induced even in relatively fragile materials such as polystyrene. Figure S11.11.1/2 shows the effect of low energy impacts on injection-moulded, weir-gated plaques of polystyrene. Plaque A is 2.6 mm thick and the cracks lie along the main flow axis. Plaque B is 6.2 mm thick and the crack pattern is more complex and consistent with a layered ordering of molecular orientation in which the skin layers have orientation similar to that predominating in the thinner mouldings, and typical of thin injection mouldings in general, and the core has an approximately transverse orientation. In such tests the impactor velocity decreases to zero and the energy density at the crack tip becomes insufficient to sustain continued crack growth.

When the low energy technique was applied to three continuous fibre composites, a sample of aromatic polymer composite (APC) and two epoxy-carbon fibre composites with the same carbon fibre content and laminate structure as the APC[1] the results showed the following:

1   Specimens 2 mm thick subjected to the standard flexed plate test (which is not specifically intended for the testing of composite materials) are damaged by incident energies between 1 and 4 joules and are seriously damaged by incident energies between 6 and 10 joules. (See also Table 11.10/1 for comparative data from excess incident energy tests.)

2   The APC was unambiguously tougher and had a higher resistance to initial damage than comparable epoxy–carbon fibre composites; the

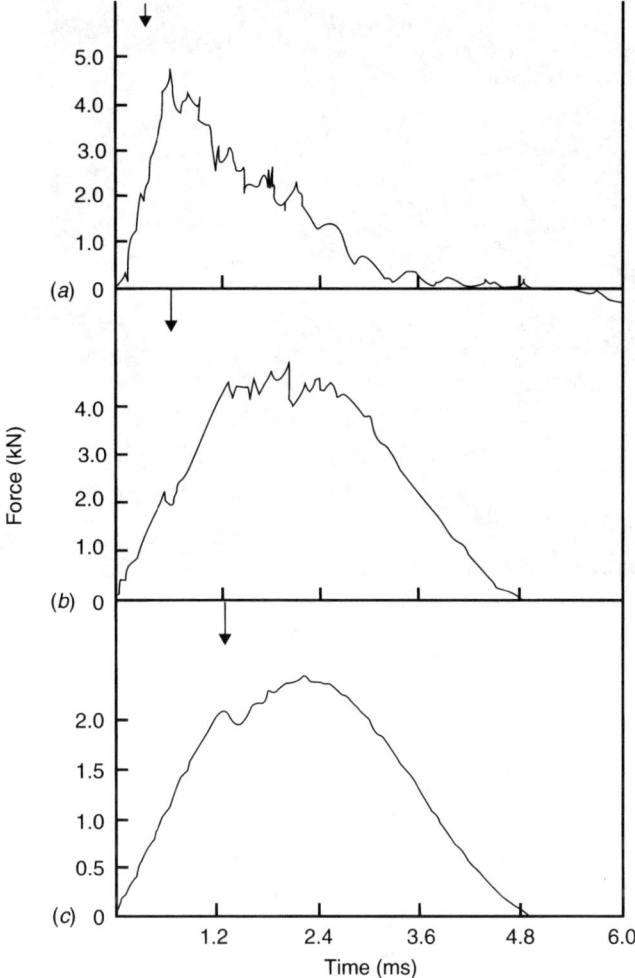

S11.11.1/1. The response of a carbon-fibre composite to impacts of different energies. Insignificant features in the response to high incident energy impact become readily discernible in the response to low incident energy impact.

relative toughnesses of the two epoxy composites reflected the relative toughnesses of the matrices.

3   The notional coefficients of restitution that were calculated from the damped vibrations were of the magnitude that would be expected and reflected the degree of damage suffered by the test pieces.

Adams and Perry,[2] amongst others, had previously used low incident energies on fibre composites and Lal[3] reported the use of coefficients of restitution in the assessment of damage to plates.

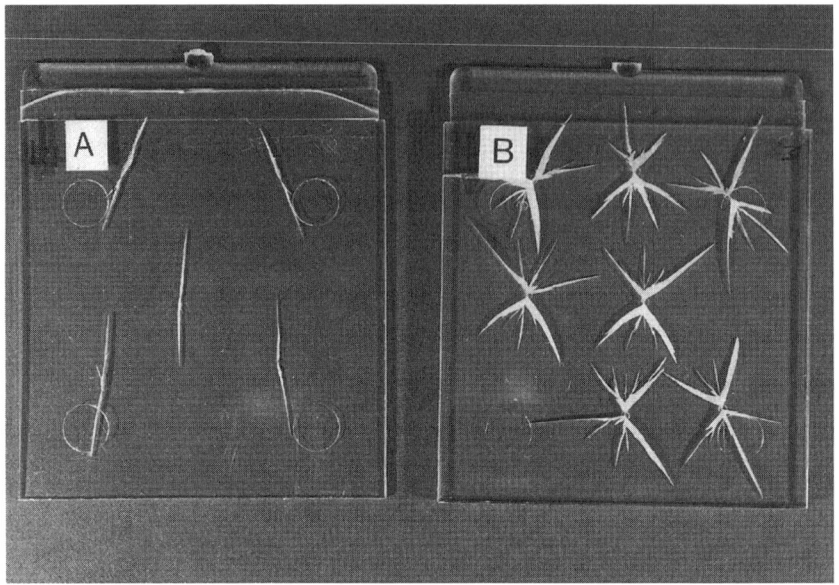

S11.11.1/2. Damage in injection mouldings of polystyrene by low
energy impacts. Plaque A is 2.6 mm thick and the cracks lies along the
main flow axes; plaque B is 6.2 mm thick and the complex crack
pattern indicates a layered ordering of the molecular orientation.

When the impactor rebounds from the test piece and strikes it again the
initial damage may be increased by the second and any subsequent impacts,
though the severity of the successive impacts diminishes progressively.
Davallo *et al.*[4] removed the possibility by a simple device that captured the
impactor after the first impact. When the test piece and the impactor vibrate
as a system, a complex modulus of the material can be calculated. For a
centrally loaded, edge-supported plate the relationship is:

$$E^* = \frac{3(1-\nu)(3+\nu)a^2 m \omega^2}{4 \pi h^3}$$

[S11.11.1.1]

where

a = radius of support ring
ω = angular frequency
h = specimen thickness
m = mass of impactor
ν = Poisson's ratio of the specimen material
E^* = the complex Young's modulus of the material

Figure S11.11.1/3 is a force–time record of a low incident energy impact
event. The impactor rebounded from the specimen, the specimen then suf-

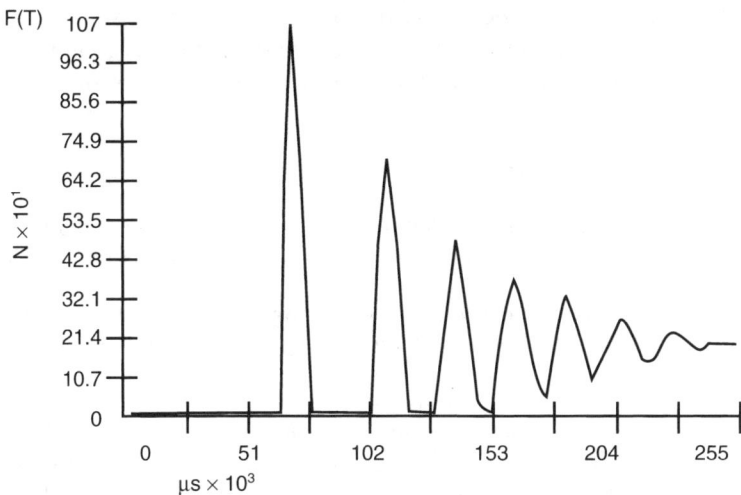

S11.11.1/3. Force–time response curve, low incident energy impact on continuous fibre composite. Depending on circumstances, the impactor may rebound from the test piece and then subject it to further impacts or may merely oscillate with it.

fered several blows of progressively diminishing severity and finally the plate–impactor combination resonated as a damped elastic system. Results obtained by this method are unlikely to be highly precise or accurate, because:

1 the stress field at and near the notional point contact is unlikely to be ideal and is never known in quantitative detail;
2 plates are seldom perfectly flat and are therefore not uniformly supported during the initial phase;
3 plates may be anisotropic and may not therefore distort symmetrically, nor will Eq. 11.11.1.1 be strictly valid;
4 slight damage to a fibre composite test piece may reduce the modulus.

The flexed beam configuration is likely to yield more accurate values of E^*.

Overall, it seems the low energy impact technique is so versatile, so informative about the train of events preceding failure and so compatible with other impact test procedures that it could be adopted as an optional variant or supplement to the standard tests. It is largely ineffective, however, for the detection of delamination damage, but since internal damage in general may be more extensive than visible damage secondary investigations, such as ultrasonic absorption or a mechanical test to destruction, can be deployed advantageously. The aircraft industry, for one, has to be concerned about damage tolerance and unsuspected damage that could subsequently induce a failure in service. A particular in-plane compression test on a plate that has been previously subjected to low incident energy impact is virtually an international standard, see reference 5 for details and comments.

References

1 P. E. Reed and S. Turner. 'Flexed plate impact, Part 7. Low energy and excess energy impacts on carbon fibre-reinforced polymer composites.' *Composites* 19, 3, 193, 1988.
2 D. F. Adams and J. L. Perry. 'Low level Charpy impact response of graphite/epoxy hybrid composites.' *Journal of Engineering Materials and Technology* 99, 257, 1977.
3 K. M. Lal. 'Coefficients of restitution for low velocity transverse impact of thin graphite–epoxy laminates.' *Composites Technology Review* 6, 3, 112, 1984.
4 M. Davallo, M. L. Clemens, H. Taylor and A. N. Wilkinson. 'Low energy impact behaviour of polyester–glass composites formed by transfer moulding.' *Plastics, Rubber and Composites: Processing and Applications* 27, 8, 384, 1998.
5 P. J. Hogg and G. A. Bibo. 'Impact and damage tolerance.' Chapter 10, '*Mechanical Testing of Advanced Fibre Composites*', ed J. M. Hodgkinson. Woodhead Publishing Limited. ISBN 1 85573 312 9.

S11.11.2 Excess energy impact

Low incident energy impact can reveal the first stages of damage and also identify the incident energy threshold below which the impacted item will survive undamaged, but the results may not correlate unambiguously with those from excess incident energy tests because survival entails the impact velocity decreasing to zero during the event. That ambiguity can be removed by, for instance, flash photography of the test piece to capture the state of damage at identifiable points during an excess energy impact event.[1] Figures S11.11.2/1 and S11.11.2/2, from reference 1, each show a photograph, one of the tensile face and one of the impacted face, of a test piece (in this case an injection moulded disc of dry nylon 66 containing 0.33 weight fraction of short glass fibres) during excess energy impact and the corresponding force–time response curve marked with the trigger point. The first photograph demonstrates that substantial damage can develop well before the peak force occurs (which can be inferred from low incident energy impact tests) and the second demonstrates that an extraneous force may persist after final disruption of the specimen. The method was subsequently consolidated[2] with improved measurement of crack length and calculations of fracture toughness (G_c). Cinephotography has been used to similar effect.

Those results and numerous similar ones are highly informative but disturbing in that they demonstrate that the usual characterizing quantities

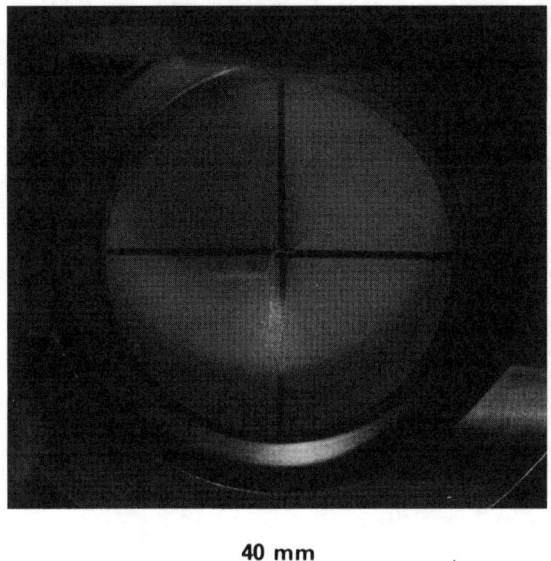

40 mm

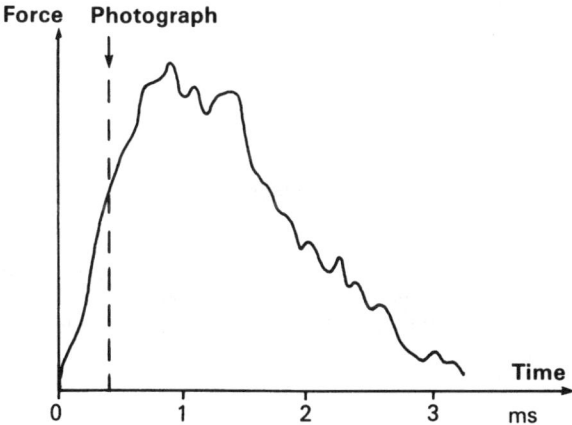

S11.11.2/1. Photograph of tensile face during impact. Glass fibre reinforced nylon. Photograph taken at absorbed energy of 0.5 joules. Force–time curve (damped) below.

peak force, energy to peak and total energy, which are nowadays easy to measure and correspondingly popular objectives in impact test programmes, are not as definitive as was widely thought when instrumented tests were first developed. On the other hand, those quantities remain

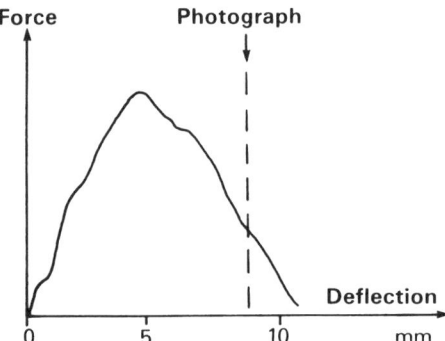

S11.11.2/2. Photograph of impacted (compression) face during impact. Glass fibre reinforced nylon. The broken test piece is being forced down into the support ring and is thereby contributing to the absorbed energy.

useful indicators of particular stages in the course of failure under impact, but primarily failure of a plastics structure, i.e. a test piece in a specific machine/test configuration, rather than failure of a plastics material. The ambiguities and physical uncertainties characteristic of the measurable quantities are probably not detrimental when the data are used for quality control and quality assurance purposes, but they are potentially misleading in the contexts of materials development, failure analysis and diagnosis. There are certain implications, also, for continuity and harmonization between modern data and those in the archives, particularly in view of the pragmatic usefulness of the latter.

References

1 A. E. Johnson, D. R. Moore, R. S. Prediger, P. E. Reed and S. Turner. 'The falling weight impact test applied to some glass-fibre reinforced nylons'. *Journal of Materials Science* 21, 3153, 1986.
2 D. R. Moore and R. S. Prediger. 'Photographed impact of continuous fiber composites.' *Polymer Engineering and Science* 28, 9, 626, 1988.

S11.14   A rationale for data generation and impact resistance databases

Various points in Chapter 11 provide a guideline on what the contents of an ideal database on impact resistance should be.

1   An impact test measures the impact resistance of that particular test piece in that particular test configuration rather than the impact resistance of the material from which the test piece is made.
2   Impact resistance is an attribute rather than a simple physical property.
3   Damage develops in an impacted object when the stress at some point exceeds the strength there.
4   The total energy absorbed during an impact is measured relatively easily but it is not simply or directly related to the energy density at the failure site which is the physically important quantity.
5   The strength varies from point to point in the impacted item through the combined effects of the processing conditions and the flow geometry.
6   Damage modifies the local stress field thereby confounding a rigorous analysis.

   The consequences are as follows:

1   The impact resistance of a material or an end-product made from it have to be inferred from that of the test pieces.
2   The conventional indicators of impact resistance, i.e. impact energy, peak force, etc. are not readily translated into an impact resistance pertinent to a particular service item, nor are they always directly linkable to a physical event.
3   Impact test data are prone to variability arising from both intrinsic and extraneous factors.
4   Data from the various established test procedures do not necessarily correlate closely nor do they necessarily correlate separately with the toughness of items impacted in service.
5   No one test method is especially favoured or has outstanding advantages except in particular, narrow contexts.

   Those consequences limit the scope of utilization of any data on impact resistance that can currently be generated. Additionally, whoever is concerned with the design of an impact resistance database has to contemplate the following range of possible inputs:

1   three basic test configurations (tension, flexed beam, flexed plate);
2   variants on specimen geometry and boundary conditions for each configuration;

3   at least five useful characterizing quantities derivable from any reputable type of impact test (each one of which is sensitive to test conditions, sample state, etc.).

Thus, the range of options for test methods and database contents is large. The options favoured by data generators are likely to depend on company or organization strategies and objectives, regional preferences, past records and present prejudices. It is highly improbable that the many currently active generators will readily adopt one uniform style for their internal databases because the various sectors of the plastics industry use properties data for different purposes. They have inevitably followed different development paths for their impact test programmes and have established particular intellectual or commercial positions which they could not change at relatively short notice and may see no advantage in ever doing so. Therefore, standardization effort should be directed at an overall format within which various standard elements can be assembled at the discretion of the compiler and in keeping with the prospective pattern of utilization.

It is clear that many matters have to be resolved by discussion before uniform and standardized data presentation styles can be established or before the information pathways between the various sectors of the industry or between data generators and data utilizers can automatically transfer pertinent data. The attainment of agreement on an international scale is always a protracted process and often tends, via the democratic process of compromise, to produce a solution that is less incisive than the ideal.

Thus, the currently recommended data fall short of what is required to characterize impact resistance over the full scope of the plastics business and even short of what is often already generated by some primary producers in the pursuance of their own development programmes. Nevertheless they constitute a major advance on what currently appears in technical brochures and public databases.

In the light of current knowledge, certain operating rules and guidelines for data generation programmes could be formulated, possibly as follows:

1   The force–time relationship should be recorded during the test and preferably converted into the force-displacement relationship (via Eqs. S11.5.1–3).
2   The basic procedure of item 1 should be carried out at whatever temperatures are relevant to the investigation but preferably at standard temperatures.
3   The test programme should embrace the processing variables, the gate geometry and the cavity geometry in the case of injection mouldings, the die geometry in the case of extrudates, etc.
4   Where appropriate, critical basic shapes/flow geometries could be tested as an integral part of the test programme.

5   Test programmes should be in balance with the main patterns of uti-
    lization of the material/prospective end-products, i.e. permissible types
    of failure, acceptable levels of damage, likelihood of impact in service,
    etc.

It is unlikely that an investigator would ever carry out a full evaluation
because of the high cost, tedium and the possibly unmanageable volume of
data that would arise. Various reduction strategies for the data generation
programmes are feasible. For instance, in view of the dominating influence
of the tough–brittle transition of thermoplastics, the impact test programme
for any specific thermoplastics sample could be reduced to one or other of
the measurable quantities at room temperature and the same quantity
measured (in possibly less detail) at three other temperatures, the lowest
well inside the brittle region; one in the transition region and the other in
the ductile region. That procedure would entail some judicious exploration
to locate the transition but that can be done by relatively cheap and expedient methods.

The stored data could consist merely of four impact property/tempera-
ture datum pairs for a reference sample, and a few supplementary data of
the same type for samples with different flow geometries or with other differences of interest to the user. Here again, provided a standard procedure
was used for each test, the limited data format would be compatible with
the overall body of knowledge. The scope of the data could be expanded
as occasion demanded.

For thermosetting plastics and long fibre composites, for which the
concept of ductile–brittle transition is inappropriate or of secondary importance, the independent variable might be cross-link density, polymer architecture, fibre volume fraction, spatial arrangement of the fibres or some
other influential factor but, as with thermoplastics, the choice of a reference
batch should be governed by the downstream pattern of usage. Apart from
such possibilities for the curtailment of test programmes, the best strategy
is for an impact test programme, or indeed any mechanical test programme,
to be so arranged that the influential variables are studied in the order of
their pertinence to the objectives of the investigation or in the order of
their increasing complexity. This can enable the planned programme to be
terminated early if the information that has already been acquired can be
deemed to satisfy the objectives of the investigation. Early termination of
a test programme detracts from the scope of an ensuing database but the
advantage will often be deemed to outweigh the disadvantage. Provided
standardized test methods rather than arbitrary ones have been used, and
provided the metadata are adequately comprehensive, the incomplete data
nevertheless become a compatible part of the expanding body of data and
knowledge about impact testing and impact resistance.

Recommended reading

Arridge, R.G.C. (1975). *Mechanics of Polymers*, Oxford University Press.

A theoretical account that links mechanical behaviour with fundamental properties.

Atkins, A.G. and Mai, Y.W. (1981). *Elastic and Plastic Fracture*, Ellis Horwood.

A broad but detailed account of fracture in materials at small and large deformations.

Brown, R.P. ed. (1981). *Handbook of Plastics Test Methods 2nd edn.* George Godwin.

A comprehensive account of test methods on many aspects of polymer properties.

Bucknall, C.B. (1977). *Toughened Plastics*, Applied Science.

An account of the means and mechanisms in toughening plastics and the consequence on their properties.

Carlsson, L.A. and Pipes, R.B. (1987). *Experimental Characterization of Advanced Composite Materials.* Prentice Hall.

Crawford, R.J. (1987). *Plastics Engineering*, Pergamon.

Mechanical and processing properties of plastics.

Folkes, M.J. (1982). *Short Fibre Reinforced Thermoplastics*, RSP.

An account of fabrication and processing of discontinuous fibre composites and their properties.

Hertzberg, R.W. and Manson, J.A. (1980). *Fatigue of Engineering Plastics*, Academic Press.

Polymer science and engineering for fatigue processes in engineering plastics.

Hodgkinson, J.M. ed. (2000). *Mechanical Testing of Advanced Fibre Composites.* Woodhead Publishing.

Kinloch, A.J. and Young, R.J. (1983). *Fracture Behaviour of Polymers*, Applied Science.

A materials science account of fracture and fracture mechanisms.

Mascia, L. (1982). *Thermoplastics: Materials Engineering.* Applied Science Publishers.

McCrum, N.G., Buckley, C.P. and Bucknall, C.B. (1997). *Principles of Polymer Engineering 2nd edn.* Oxford University Press.

An introductory text to broad aspects of polymer mechanical properties.

Moore, D.R., Pavan, A. and Williams J.G. eds. (2001). *Experimental Methods in the Application of Fracture Mechanics Principles to the Testing of Polymers, Adhesives and Composites,* Elsevier.

A multiauthor book containing test protocols and background science for conducting fracture mechanics tests.

Nielsen, L.E. (1974). *Mechanical Properties of Polymers and Composites,* Marcel Dekker.

Two-volume text on the general principles of the mechanical behaviour of polymers and composites.

Paul, D.R. and Bucknall, C.B. eds. (2000). *Polymer Blends Vol 2-Perfomance,* John Wiley.

A multiauthor book on wide aspects of the properties and mechanisms of deformation in polymer blends.

Pipes, R.B. ed. (1991). *Composite Materials Series,* Elsevier.

A six part series on topics including friction and wear, fibre reinforcement, textile structure, fatigue, interlaminar response and application of fracture mechanics.

Roark, R.J. and Young, W.C. (1975). *Formulas for Stress and Strain,* McGraw-Hill Kogakusha.

A condensed summary of linear elastic stress theory in the form of helpful formulae.

Struik, L.C.E. (1977). *Physical Aging in Amorphous Polymers and Other Materials,* TNO.

A fundamental account of ageing effects in polymers.

Stuart, H.A. ed. (1956). *Die Physik der Hochpolymeren,* Springer-Verlag.

A multiauthor book containing much of the theoretical basis for linear viscoelasticity.

Swallowe, G.M. ed. (1999). *Mechanical Properties and Testing of Polymers,* Kluwer Academic Publishers.

A multiauthor book with many small articles on the science and technology of mechanical properties and polymer testing.

Treloar, L.R.G. (1975). *The Physics of Rubber Elasticity 3rd edn.* Oxford University Press.

Comprehensive text on the mechanical properties of rubbers.

Ward, I.M. (1983). *Mechanical Properties of Solid Polymers 2nd edn.* Wiley.

Comprehensive account of the physics of mechanical properties.

Whitney, J.M., Daniel, I.M. and Pipes, R.B. (1982). *Experimental Mechanics of Fiber Reinforced Composite Materials. SESA Monograph No. 4.* Society for Experimental Stress Analysis.

Williams, J.G. (1984). *Fracture Mechanics of Polymers*, Ellis Horwood.

A fundamental account of the application of fracture mechanics principles to polymers.

Williams, J.G. and Pavan, A. eds. (1995). *Impact and Dynamic Fracture of Polymers and Composites,* MEP.

A contemporary set of papers of key issues in impact and high speed fracture.

Williams, J.G. and Pavan, A. eds. (2000). *Fracture of Polymers, Composites and Adhesives,* Elsevier.

A contemporary set of papers setting out the state of the art in fracture of polymers, composites and adhesives.

Wright, D.C. (1996). *Environmental Stress Cracking of Plastics.* RAPRA Technology Ltd.

Index